James B. Thomson

Illustrated Table-Book

Or, Juvenile arithmetic, containing oral and slate exercises for beginners

James B. Thomson

Illustrated Table-Book
Or, Juvenile arithmetic, containing oral and slate exercises for beginners

ISBN/EAN: 9783337384203

Printed in Europe, USA, Canada, Australia, Japan

Cover: Foto ©berggeist007 / pixelio.de

More available books at **www.hansebooks.com**

THOMSON'S NEW MATHEMATICAL SERIES.

ILLUSTRATED

TABLE-BOOK,

OR

JUVENILE ARITHMETIC,

CONTAINING

ORAL AND SLATE EXERCISES

FOR

BEGINNERS.

By JAMES B. THOMSON, LL.D.,

AUTHOR OF NEW GRADED SERIES OF ARITHMETICS, PRACTICAL ALGEBRA, ETC.

NEW YORK:
CLARK & MAYNARD, PUBLISHERS,
5 BARCLAY STREET,
1889.

THOMSON'S MATHEMATICAL SERIES.

I. *A Graded Series of Arithmetics, in three Books, viz.:*

New Illustrated Table Book, or Juvenile Arithmetic. With oral and slate exercises. (For beginners.) 128 pp.

New Rudiments of Arithmetic. Combining Mental with Written Arithmetic. (For Intermediate Classes.) 224 pp.

New Practical Arithmetic. Adapted to a complete business education. (For Grammar Departments.) 384 pp.

II. *Independent Books.*

Key to New Practical Arithmetic. Containing many valuable suggestions. (For teachers only) 168 pp.

New Mental Arithmetic. Containing the Simple and Compound Tables. (For Primary Schools.) 144 pp.

Complete Intellectual Arithmetic. Specially adapted to Classes in Grammar Schools and Academies. 168 pp.

III. *Supplementary Course.*

New Practical Algebra. Adapted to High Schools and Academies. 312 pp.

Key to New Practical Algebra. With full solutions. (For teachers only.) 224 pp.

New Collegiate Algebra.

Complete Higher Arithmetic. (In preparation)

* *Each book of the Series is complete in itself.*

PREFACE.

In order to pursue the study of Arithmetic with pleasure and success, two things are essential. First, the *elementary* principles and *Tables* must be *thoroughly understood;* Second, their application must be *made perfectly familiar.*

The *Revised Table Book* is designed to assist both the teacher and the pupil in the accomplishment of these important ends. It is constructed on the following plan:

1. As soon as a child learns a fact or principle in arithmetic, he is taught its *application,* and begins to *practice* it. In this way an interest is awakened in the subject, and the pupil is relieved from the irksomeness of learning and repeating abstract results and principles, while *ignorant* of their *nature* and *use.*

2. The pupil is taught to *illustrate* and *form* the *Simple Tables* for himself. A child gains a much clearer idea of a thing by doing it himself, than by having his teacher do it for him.

3. Alternate Mental and Slate Exercises are interspersed throughout the book. These exercises are carefully adapted to the capacity of beginners, and to the principles they are designed to illustrate. They should therefore be made perfectly familiar by *personal application* either at home or at school. In this way the pupil will learn how to *study,* how to *think,* and how to *reason.*

4. Special pains have been taken to correct the *Tables* of *Weights and Measures,* according to present Law and usage; the obsolete denominations and those not used in the United States being carefully excluded.

APPARATUS.—Beginners in Arithmetic should be furnished with suitable tools; as slates, blackboards, numeral frames, counters, " counting boards," etc., and *be taught to use* them in learning to count, and in illustrating the elementary combinations and principles of numbers.

Movable objects used for counters, or *unit marks* made by the pupil upon a slate or blackboard, are preferable to the *fixed objects* seen in pictures, or the stationary numbers represented by figures, straight lines, or other unit marks upon charts and wall maps.

The COUNTING BOARD is a new and valuable accession to the apparatus of primary schools. Its length depends upon the size of the recitation room, and the number of pupils in the class. When placed against the wall, its width is usually from 15 to 18 inches; the upper surface is divided into parts by distinct marks or strips of wood; the edges are faced with a plain moulding, raised sufficiently to prevent the counters from rolling off. If placed in the middle of the room, it should be twice this width, to allow pupils to stand on either side of it. (*See Cover.*)

JAMES B. THOMSON.

NEW YORK, August, 1874.

NOTE.

THE few changes in the fore part of this Edition of the *Revised Table-Book,* will not prevent its use in connection with the former editions.

November, 1877.

COUNTING.

LESSON I.

1. How many of these little girls and boys wish to learn to count? All that do may hold up a hand.

2. How many hands does each hold up?
One.
3. How many are one hand and one hand?
Two.
4. How many hands have you? How many feet?
5. How many are two pencils and one pencil?
Three.

6. Show me three fingers and count them.

7. Which is your right hand? Which your left?

8. How many thumbs have you on your right hand? How many on your left hand?

9. How many on both?

10. How many are two fingers and one finger?

11. Three is how many more than two?

LESSON II.

To Teachers.—A child learns much faster by *doing* or illustrating a thing, than by simply *repeating* an explanation which he hears, but does not understand. Hence, the pupil at every exercise, should be called upon to do as much as possible with his own hands. This is the only way to insure rapid progress and a thorough knowledge of numbers.

1. Let each show me two fingers. Three fingers.

2. Clap your hands three times in concert?

3. What comes next after one? After two?

4. Three is how many more than one? Show it with your fingers.

5. Count three, beating time with your right hand.

6. Count from three back to one.

7. Each show three fingers on the left hand? Now the other finger.

8. How many are three fingers and one finger? *Four.*

9. How many are three pencils and one pencil?

10. Clap your hands four times in concert?

11. What comes next after two? After three?

12. How many more than three is four? Than one?

13. What comes next before four? Before three?

14. Count from four back to one.

15. How many fingers on your left hand? "Four."
16. Count your thumb with them; how many?
Five.
17. Count five in concert, beating time.
18. Count from five back to one. Twice more.

LESSON III.

1. When we say one, two, three, four, &c., what is it called?
Counting.
2. Count five in concert, beating time.
3. Rap on your slate with your pencil five times.
4. Copy the figures used to express the first five numbers, as I make them upon the blackboard.

| 1, | 2, | 3, | 4, | 5. |
| one, | two, | three, | four, | five. |

5. Five is how many more than four? Show it.
6. Show by your fingers how many things are denoted by the figure 3. By 2. By 4. By 5.

7. George may make five straight marks upon the blackboard; the others, on their slates.

8. Count them, pointing to each.

9. Make another; how many have you now?
Six.

10. Count six in concert, beating time.

11. Show me six fingers.

12. Make six raps on your slate with your pencil.

13. Six is how many more than five?   Show it?

14. Six is how many more than one?

15. What comes next before six?   Next before five? Before four?   Before three?   Before two?

16. Count from six back to one.

LESSON IV.

To Teachers.—The division into lessons is not for the purpose of fixing the daily task of pupils, but to point out the groups of numbers which should be taught consecutively. It may be necessary to subdivide some of them into several recitations. But however this may be, the class should not be permitted to pass from one to the next, till the former is thoroughly understood.

1. Here is a pile of six books; if I put another with them, how many will there be?
Seven.

2. Count seven in concert, beating time.

3. Who can count seven alone promptly?   Henry may try.   Carrie.   William.   Louise.

4. Let each show me seven fingers.

5. Count from seven back to one.

6. Seven is how many more than six?

7. Show it by counters.

8. Seven is how many more than one?

9. Make seven unit marks upon your slates.

10. How many are seven pencils and one pencil?

Eight.

11. Count eight in concert, beating time.

12. Let each count eight alone. (The teacher names individuals.)

13. Make eight raps on your slate with your pencil?

14. Eight is how many more than seven?

15. Show it by counters.

16. Eight is how many more than one?

17. Count from eight back to one.

18. How many are eight counters and one counter?

Nine.

19. Count nine in concert, beating time.

20. Show me nine objects.

21. What comes next before nine? Next before eight? Before seven? Before six? Before five? Before four? Before three?

22. Count from five to nine, inclusive.

23. How many are nine cents and one cent more?

Ten.

24. How many fingers and thumbs has each?

25. Count ten, beating time?

26. Write the figures denoting six, seven, eight, nine, ten.

6,	7,	8,	9,	10.
six,	seven,	eight,	nine,	ten.

27. Show by marks upon your slate how many things are denoted by the figure 6. By the figure 8. By the figure 7. By the figure 9.

LESSON V.

1. Count the balls on the upper wire of the Numeral Frame, as I move them from left to right. How many?

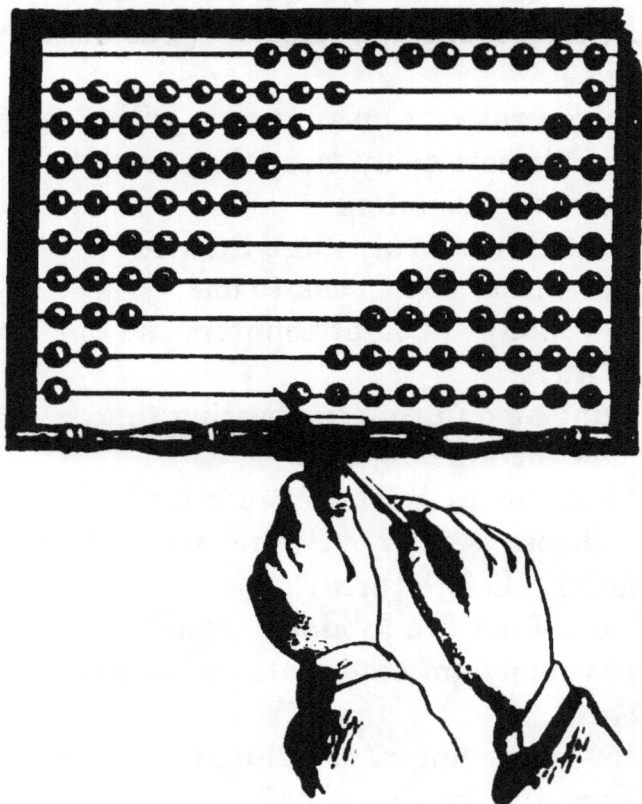

2. If I move across one ball on the next wire, how many will it make counted with ten? *Eleven.*
3. If I move two on the next, how many? *Twelve.*
4. If three balls on the next, how many? *Thirteen.*
5. If four balls on the next, how many? *Fourteen.*
6. If five balls on the next, how many? *Fifteen.*
7. If six balls on the next, how many? *Sixteen.*

8. If seven balls on the next, how many? *Seventeen.*
9. If eight balls on the next, how many? *Eighteen.*
10. If nine balls on the next, how many? *Nineteen.*
11. If ten balls on the next, how many? *Twenty.*

LESSON VI.

1. What comes next after ten? "Eleven."
2. Eleven is how many more than ten? "One."
3. How represent eleven by the Numeral Frame?
Ans. By moving across ten balls on the upper wire and one on the next.
4. What comes next after eleven?
5. Twelve is how many more than eleven?
6. How represent twelve by the Numeral Frame?
Ans. By the ten balls on the upper wire and two on another.
7. What comes next after twelve?
8. Thirteen is how many more than twelve? Than ten?
9. How show that thirteen is three more than ten?
Ans. By the ten balls on the upper wire and three on another.
10. The meaning of the word thirteen?
Ans. Three and ten.
11. What comes next after thirteen?
12. The meaning of the word fourteen?
Ans. Four and ten.
13. How show that fourteen is four more than ten?
14. What comes next after fourteen?
15. The meaning of the word fifteen?
Ans. Five and ten.

16. How show that fifteen is five more than ten?

17. What comes next after fifteen?

18. The meaning of the word sixteen?
    Ans. Six and ten.

19. How show that sixteen is six more than ten?

20. What comes next after sixteen? After seventeen?

21. Next after eighteen? After nineteen?

22. Count from ten to twenty, beating time.

23. What is the meaning of the word twenty?
    Ans. Two tens.

24. Write the figures denoting eleven, &c., to fifteen.

      11,   12,   13, . 14,   15.
    eleven,   twelve,   thirteen, fourteen, fifteen.

25. What comes next after 13? After 11? After 14? After 12?

26. Which is the greater, fifteen or thirteen? How show it?

27. Write the figures denoting sixteen, &c., to twenty.

      16,   17,   18,   19,   20.
    sixteen, seventeen, eighteen, nineteen, twenty.

28. Which is the greater, seventeen or fifteen? Show it.

29. Which is the greater, nineteen or eighteen?

30. Write in figures thirteen, seventeen, twelve, ten, eighteen, sixteen, eleven, twenty.

☞ Having learned to count ten, children will easily learn to count from ten to twenty, by observing that after twelve, the names of the successive numbers are formed by prefixing the names of the numbers, three, four, five, six, &c., in their order to *teen* or *ten*. They will also find great assistance in comprehending the meaning of the terms from ten to a hundred by directing attention to their derivation.

LESSON VII.

1. What does the word twenty denote?
Ans. Twenty denotes *two tens.*
2. How is twenty written?
Ans. By writing 2 in the second place, with a cipher on the right; as, 20.
3. Count the marks, as I make them upon the blackboard; how many? "Ten."
4. Count on, as I make another row. "Twenty."
5. If I make another, how many? "Twenty-one."
6. If I make another, how many? "Twenty-two."
7. If another, how many? "Twenty-three."
8. If another, how many? "Twenty-four."
9. If another, how many? "Twenty-five."
10. If another, how many? "Twenty-six."
11. If another, how many? "Twenty-seven."
12. If another, how many? "Twenty-eight."
13. If another, how many? "Twenty-nine."
14. If another, how many? "Thirty."
15. Twenty-one is how many more than twenty?
16. Twenty-two is how many more than twenty?
17. Twenty-five, than twenty? Show it.
18. Twenty-seven, than twenty? Show it.
19. Thirty is how many more than twenty? Show it.
20. Count from twenty to thirty in concert.
21. Write the figures denoting twenty-one, twenty-two, &c., to thirty.
22. What do the figures 23 standing side by side denote? What 25? What 27? What 26? What 29? What 28? What 22?

LESSON VIII.

1. What is the meaning of the word thirty?
Ans. Thirty denotes *three tens.*

2. How is thirty expressed?
Ans. By writing 3 in the second place with a cipher on the right; as, 30.

3. Count from thirty to forty.

4. Write in figures thirty-one, &c., to forty.

5. Count from forty to fifty, in like manner.

6. Write in figures from forty to fifty.

7. What do the figures 25, denote? The figures 34? The figures 37? 43? 48? 50?

8. Count from fifty to sixty, beating time?

9. Make the figures denoting fifty-three? Fifty-seven? Fifty-five? Fifty-eight?

10. What do the figures 52, denote? 54? 56? 59?

11. Count from sixty to seventy, in like manner.

12. Write the figures from sixty to seventy?

13. Count from seventy to eighty?

14. Make the figures denoting sixty-six. Seventy-three. Seventy-five. Seventy-nine. Eighty.

15. Count from eighty to ninety.

16. Write the figures from eighty to ninety.

17. Count from ninety to a hundred.

18. Write the figures from ninety to a hundred.

19. What do the figures 12 denote? What 21?
Ans. The figures 12, denote 1 ten and 2 units; 21 denotes 2 tens and 1 unit.

20. What do the figures 34 denote? 47? 63? 75? 88? 93? 100?

LESSON IX.

To Teachers.—As soon as children have learned to count and write one hundred with accuracy, they will find no difficulty in counting and writing larger numbers.

1. Count from one hundred to one hundred and ten.
2. Write the figures from one hundred to one hundred and ten.

100, 101, 102, 103, 104, 105, 106, 107, 108, 109

3. How is one hundred and ten expressed?

Ans. By writing 1 in the third place, 1 in the second, and a cipher in the first; as, 110.

4. Count from one hundred and ten to one hundred and twenty.

"One hundred eleven, one hundred twelve, &c."

5. Write in figures from 110 to one hundred twenty.

111, 112, 113, 114, 115, 116, 117, 118, 119, 120.

6. Count from one hundred twenty to one hundred thirty, in like manner.

7. Count from one hundred thirty to one hundred forty, and so on to two hundred.

8. Write in figures one hundred thirty. One thirty-seven. One thirty-nine.

9. Write in figures each decade from 140 to 200.

10. What do the figures 135, denote?

Ans. One hundred, 3 tens, and 5 units.

11. What do the figures 124, denote? 136? 140? 159?

12. What do the figures 160, denote? 177? 185? 188? 190? 199? 200?

LESSON X.

1. How is two hundred expressed by figures?

Ans. By writing 2 in the third place, with two ciphers on the right; as, 200.

2. How express three hundred? Four hundred, &c., to nine hundred?

3. Write in figures two hundred and ten. Two hundred and twenty, &c., to three hundred.

4. Write three hundred twenty-five. Three hundred forty-six. Three hundred fifty-four.

5. Write five hundred twenty-seven. Five thirty-eight. Five forty-six. Five sixty-nine.

6. Write six hundred thirty-three. Six forty-eight. Six fifty-two. Seven ninety-four.

7. Write nine hundred fifteen. Nine twenty-eight. Nine forty-three. Nine sixty. Nine ninety-nine.

8. What do the figures 235, denote?

Ans. 2 hundreds, 3 tens, and 5 units.

9. What do the figures in 346, denote? In 378? Copy and read the following numbers:

1. 408.	10. 525.	19. 703.
2. 310.	11. 405.	20. 890.
3. 101.	12. 239.	21. 875.
4. 230.	13. 350.	22. 940.
5. 175.	14. 475.	23. 963.
6. 240.	15. 509.	24. 893.
7. 301.	16. 643.	25. ·985.
8. 2ʰ	17. 538.	26. 990.
9. 305.	18. 720.	27. 1000.

(For explanation of terms, see p. 40.)

ADDITION TABLES.

LESSON I.

To Teachers.—The present Lesson consists in adding the unit *one* to the several digits in succession. As the results correspond with the regular increase of numbers by counting, they need no further illustration. The object of the next eight lessons is to familiarize the learner with the addition of the other digits to each other.

0	and	1	are	1.	5	and	1	are	6.
1	"	1	"	2.	6	"	1	"	7.
2	"	1	"	3.	7	"	1	"	8.
3	"	1	"	4.	8	"	1	"	9.
4	"	1	"	5.	9	"	1	"	10.

1. If you have three pears and I give you one pear, how many pears will you have?

Solution.—Three pears and 1 pear are 4 pears.

2. How many are 4 apples and 1 apple? Show it with your fingers.

3. How many are 5 pencils and 1 pencil? Show it.

4. George had 6 cents and earned 1 more: how many had he then?

5. How many are 7 marbles and 1 marble? Show it.

6. Eight tops and 1 top are how many? Show it.

7. If a teacher has 9 roses and a pupil gives her 1 more, how many roses will she have?

8. Write the following numbers in figures: Thirteen, eleven, nineteen, twenty, twenty-five, thirty, and forty.

2

LESSON II.

To Teachers.—The Addition Table of 2 may be illustrated thus:

✳ and ✳ ✳ are *three.*

✳ ✳ and ✳ ✳ " *four.*

✳ ✳ ✳ and ✳ ✳ " *five.*

✳ ✳ ✳ ✳ and ✳ ✳ " *six,* &c.

Let the class finish the illustration, writing the table out in full.

0	and	2	are	2.	5 and 2 are	7.
1	"	2	"	3.	6 " 2 "	8.
2	"	2	"	4.	7 " 2 "	9.
3	"	2	"	5.	8 " 2 "	10.
4	"	2	"	6.	9 " 2 "	11.

1. If George has 4 pencils and buys 2 more, how many pencils will he have?

Solution.—Four pencils and 2 pencils are 6 pencils.

2. How many are 3 slates and 2 slates? Show it.
3. How many are 5 hats and 2 hats? Show it?
4. How many are 6 roses and 2 roses?
5. How many are 8 cherries and 2 cherries?
6. Add 2 to itself continually, till the sum is 50.

Copy and add the following:

(7.)	(8.)	(9.)	(10.)	(11.)	(12.)	(13.)	(14.)
1	1	1	2	2	1	2	1
2	2	2	1	1	2	1	2
3	4	6	5	7	2	8	9

Note.—As soon as the combinations of the digits are understood, they should be so thoroughly fixed in the memory, that when any result is required, it should instantly flash upon the mind.

LESSON III.

0	and	3	are	3.		5	and	3	are	8.
1	"	3	"	4.		6	"	3	"	9.
2	"	3	"	5.		7	"	3	"	10.
3	"	3	"	6.		8	"	3	"	11.
4	"	3	"	7.		9	"	3	"	12.

1. How many are 4 boys and 3 boys? Show it.

2. How many are 5 needles and 3 needles? Show it.

3. George has 3 flowers in one hand and 6 in the other: how many has he in both hands?

4. 3 cents and 9 cents are how many cents?

5. 7 knives and 3 knives are how many?

6. If 8 chickens are under the coop and 3 outside of it, how many chickens are there in all?

7. Harry's kite line is 9 yards long: if he ties on 3 yards more, how long will it then be?

8. Add 3 to itself continually, till the sum is 60.

9. What is uniting two or more numbers in one, called?

Addition.

10. What is the result or number obtained called?

The *Sum* or *Amount.*

11. When we say 2 and 1 are 3 and 2 are 5, which is the sum?

Copy and add the following:

(12.)	(13.)	(14.)	(15.)	(16.)	(17.)	(18.)	(19.)
2	1	2	3	2	2	2	3
3	3	3	3	3	3	1	1
4	5	2	4	6	7	8	9

LESSON IV.

O	and	4	are	4.	5	and	4	are	9.
I	"	4	"	5.	6	"	4	"	10.
2	"	4	"	6.	7	"	4	"	11.
3	"	4	"	7.	8	"	4	"	12.
4	"	4	"	8.	9	"	4	"	13.

1. How many are 4 grapes and 5 grapes?

2. Charles has 4 credits: how many must he get to have 7?

3. 4 and what number are 6?

4. If you have 6 rabbits and buy 4 more, how many will you then have?

5. How many are 9 cents and 4 cents?

6. 7 ducks and 4 ducks are how many?

7. William has 4 apples and George has 4 more than William: how many has George?

8. How many are 8 girls and 4 girls?

9. Julia has 4 peaches and Fanny has 3 more than Julia; how many has Fanny?

10. How many are 9 tops and 4 tops?

11. Add 4 to itself continually till the sum is 60.

Copy and add the following:

(12.)	(13.)	(14.)	(15.)	(16.)	(17.)	(18.)	(19.)
1	2	2	3	4	3	2	3
2	1	4	3	2	1	4	4
3	4	3	2	1	4	1	8
4	3	2	4	2	3	4	2
5	4	3	6	8	7	5	9

LESSON V.

0	and	5	are	5.	5	and	5	are	10.
1	"	5	"	6.	6	"	5	"	11.
2	"	5	"	7.	7	"	5	"	12.
3	"	5	"	8.	8	"	5	"	13.
4	"	5	"	9.	9	"	5	"	14.

1. 5 pencils and 3 pencils are how many?

2. 5 apples and 4 apples are how many?

3. There are 5 birds on one tree and 5 on another: how many birds are on both trees?

4. On one rose-bush are 6 buds, and on another 5: how many buds are on both?

5. How many are 5 trunks and 7 trunks?

6. How many are 8 guns and 5 guns?

7. Moses wrote 9 lines and Margaret 5 lines: how many lines did both write?

8. Add 5 to itself continually, till the sum is 60.

9. How many are 11 and 5? 12 and 5? 14 and 5? 13 and 5? 16 and 5? 15 and 5? 18 and 5? 17 and 5? 19 and 5?

10. Write the following in figures: Forty-nine, Fifty-two, Sixty-seven, Eighty-one, Seventy-four, Eighty-six, One hundred and fifteen.

Copy and add the following:

(11.)	(12.)	(13.)	(14.)	(15.)	(16.)	(17.)	(18.)
2	2	4	2	4	4	2	4
3	1	2	3	3	1	4	2
2	3	3	5	1	2	5	3
5	5	5	4	2	5	3	5
4	4	3	2	5	3	4	2

LESSON VI.

0 and 6 are 6.	5 and 6 are 11.
1 " 6 " 7.	6 " 6 " 12.
2 " 6 " 8.	7 " 6 " 13.
3 " 6 " 9.	8 " 6 " 14.
4 " 6 " 10.	9 " 6 " 15.

1. How many are 5 dollars and 6 dollars?

2. If you have 6 pears and buy 4 more, how many will you then have?

3. How many are 10 melons and 6 melons?

4. If you pick 6 peaches from one tree, and 7 from another, how many will you pick from both?

5. If there are 8 birds on a tree, and 6 more on the ground, how many are there in all?

6. George paid 6 cents for a ball, and 6 cents for an orange: how many cents did he pay for both?

7. Add 6 to itself continually, till the sum is 60.

8. Express Forty-four in figures, Fifty, Sixty-five, Seventy-eight, Eighty-five, Ninety-three.

9. How many are 11 and 6? 13 and 6? 14 and 6? 12 and 6? 15 and 6? 17 and 6? 16 and 6? 18 and 6? 19 and 6?

Copy and add the following:

(10.)	(11.)	(12.)	(13.)	(14.)	(15.)	(16.)	(17.)	(18.)
2	5	4	3	2	4	2	5	3
4	2	3	5	4	3	4	4	5
5	3	2	6	3	6	3	6	4
6	6	6	4	6	5	6	3	6
3	4	5	2	4	3	2	4	5

LESSON VII.

0	and	7	are	7.	5	and	7	are 12.
1	"	7	"	8.	6	"	7	" 13.
2	"	7	"	9.	7	"	7	" 14.
3	"	7	"	10.	8	"	7	" 15.
4	"	7	"	11.	9	"	7	" 16.

1. Sarah has 3 dolls, and Louise has 7 : how many have both?

2. 5 hats and 7 hats are how many? 6 and 7?

3. George spent 4 cents for marbles, and 7 cents for a sponge : how many cents did he spend in all?

4. How many are 7 dollars and 5 dollars? 8 and 7?

5. Charles obtained 12 credits in the morning, and 7 in the afternoon : how many had he in both?

6. How many are 7 caps and 9 caps?

7. George has 8 doves, and has sold 7 : how many had he at first?

8. Add 7 to itself till the sum is 70.

9. Express in figures, Fifty-nine, Seventy-eight, Eighty-four, Sixty-two, Ninety-three, Ninety-seven.

10. How many are 11 and 7? 12 and 7? 14 and 7? 13 and 7? 15 and 7? 18 and 7? 19 and 7?

Copy and add the following:

(11.)	(12.)	(13.)	(14.)	(15.)	(16.)	(17.)	(18.)
5	4	2	3	5	4	5	6
4	2	3	5	4	5	3	5
2	3	7	3	3	2	6	4
3	7	6	4	2	3	7	5
7	6	5	7	6	7	2	6
2	3	4	3	4	5	4	7

LESSON VIII.

0	and	8	are	8.	5	and	8	are	13.
1	"	8	"	9.	6	"	8	"	14.
2	"	8	"	10.	7	"	8	"	15.
3	"	8	"	11.	8	"	8	"	16.
4	"	8	"	12.	9	"	8	"	17.

1. Jane had 2 needles on her work, and 8 in her needlebook : how many had she in all?

2. Julia paid 8 cents for a thimble, and 5 cents for a spool of cotton : how much did she pay for both?

3. How many are 7 yards and 8 yards? 9 and 8?

4. Edward bought figs for 6 cents, and cakes for 8 cents : how much did he pay for both?

5. How many are 4 feet and 8 feet? 3 and 8?

6. One pupil gave the teacher 7 pinks, and another gave 8 : how many did both give her?

7. Add 8 to itself till the sum is 80.

8. Write in figures, Fifty-four, Sixty-four, Seventy-five, Sixty-nine, Eighty-eight, Ninety-six, a Hundred.

9. How many are 11 and 8? 13 and 8? 12 and 8? 14 and 8? 16 and 8? 15 and 8? 17 and 8? 19 and 8?

Copy and add the following:

(10.)	(11.)	(12.)	(13.)	(14.)	(15.)	(16.)	(17.)
6	5	6	2	4	5	6	6
5	4	2	3	3	3	7	5
3	2	3	5	5	8	4	7
4	3	8	8	2	5	6	6
8	8	4	5	8	7	8	8
3	4	5	7	6	4	6	7

LESSON IX.

0	and	9	are	9.		5	and	9	are	14.
1	"	9	"	10.		6	"	9	"	15.
2	"	9	"	11.		7	"	9	"	16.
3	"	9	"	12.		8	"	9	"	17.
4	"	9	"	13.		9	"	9	"	18.

NOTE.—The first five numbers are easily added. The results of adding 9, being 1 less than if 10 were added, are also easily remembered. The others, 6, 7, 8, are more difficult, and therefore should receive *special attention*.

1. Charlotte solved 4 examples in Addition, and 9 in Subtraction: how many did she solve in all?

2. James has 3 pears, and gave away 9: how many had he at first?

3. How many are 6 books and 9 books? 9 and 7?

4. 9 dollars and 5 dollars are how many dollars?

5. Joseph played 5 games of checkers, and 9 games of dominoes: how many games did he play in all?

6. William is 8 years old: how old will he be 9 years hence?

7. Helen has given away 9 apples, and has 6 left: how many had she at first?

8. Add 9 to itself, till the sum is 90.

Copy and add the following:

(9.)	(10.)	(11.)	(12.)	(13.)	(14.)	(15.)	(16.)
8	5	8	5	6	8	7	9
7	9	5	6	5	7	8	6
7	5	4	7	7	5	7	8
5	8	6	9	3	9	6	7
9	7	9	4	9	7	8	9
4	6	3	7	8	6	7	8

LESSON X.

Oral Drill.

To Teachers.—The object of this and other *Oral Drills*, is to secure accurate and rapid combinations. Each example should be continued through the ten decades, and be dwelt upon till perfectly familiar.

1. How many are 2 and 10? 12 and 10? 22 and 10? 32 and 10? 42 and 10? etc., to 92.
2. 2 and 2? 12 and 2? 22 and 2 ? etc., to 92.
3. 1 and 3 ? 11 and 3 ? 21 and 3, etc., to 94 ?
4. 2 and 3 ? 12 and 3 ? 22 and 3, etc., to 95 ?
5. 3 and 3 ? 13 and 3 ? 23 and 3, etc., to 96 ?
6. 1 and 4 ? 11 and 4 ? 21 and 4 ? 31 and 4 ?
7. 2 and 4 ? 12 and 4 ? 22 and 4 ? 32 and 4 ?
8. 3 and 4 ? 13 and 4 ? 23 and 4 ? 33 and 4 ?
9. 4 and 4 ? 14 and 4 ? 24 and 4 ? 34 and 4 ?
10. 3 and 10 ? 13 and 10 ? 23 and 10 ? etc.
11. 4 and 10 ? 14 and 10 ? 24 and 10 ? etc.
12. 5 and 10 ? 15 and 10 ? 25 and 10 ? etc.
13. 6 and 10 ? 16 and 10 ? 26 and 10 ? etc.
14. 7 and 10 ? 17 and 10 ? 27 and 10 ? etc.
15. 8 and 10 ? 18 and 10 ? 28 and 10 ? etc.
16. 2 and 5 ? 12 and 5 ? 22 and 5 ? 32 and 5 ? etc.
17. 3 and 5 ? 13 and 5 ? 23 and 5 ? 33 and 5 ? etc.
18. 4 and 5 ? 14 and 5 ? 24 and 5 ? 34 and 5 ? etc.
19. 5 and 5 ? 15 and 5 ? 25 and 5 ? 35 and 5 ? etc.

☞ The learner will observe that the right hand figure of the sum of two digits in every decade, is always the same. Thus, 3 and 4 are 7 ; 13 and 4 are 17 ; 23 and 4 are 27, etc., the right hand figure being always 7.

LESSON XI.

1. How many are 2 and 6? 12 and 6? 22 and 6? 32 and 6? 42 and 6? 52 and 6? 62 and 6? 72 and 6? 82 and 6? 92 and 6?
2. 3 and 6? 13 and 6? 23 and 6? 33 and 6? etc.
3. 4 and 6? 14 and 6? 24 and 6? 34 and 6? etc.
4. 5 and 6? 15 and 6? 25 and 6? 35 and 6? etc.
5. 6 and 6? 16 and 6? 26 and 6? 36 and 6? etc.
6. 1 and 5? 11 and 7? 21 and 7? 31 and 7? etc.
7. 2 and 7? 12 and 7? 22 and 7? 32 and 7? etc.
8. 3 and 7? 13 and 7? 23 and 7? 33 and 7? etc.
9. 4 and 7? 14 and 7? 24 and 7? 34 and 7? etc.
10. 5 and 7? 15 and 7? 25 and 7? 35 and 7? etc.
11. 6 and 7? 16 and 7? 26 and 7? 36 and 7? etc.
12. 7 and 7? 17 and 7? 27 and 7? 47 and 7? etc.
13. 11 and 8? 21 and 8? 31 and 8? 41 and 8? etc.
14. 2 and 8? 12 and 8? 22 and 8? 32 and 8? etc.
15. 3 and 8? 13 and 8? 23 and 8? 33 and 8? etc.
16. 4 and 8? 14 and 8? 24 and 8? 34 and 8? etc.
17. 5 and 8? 15 and 8? 25 and 8? 35 and 8? etc.
18. 6 and 8? 16 and 8? 26 and 8? 36 and 8? etc.
19. 7 and 8? 17 and 8? 27 and 8? 37 and 8? etc.
20. 8 and 8? 18 and 8? 28 and 8? 38 and 8? etc.

Write the following numbers in figures:

21. Fifty-seven, Sixty-eight, Eighty, Ninety-three, One hundred and one, One hundred and ten.

Copy and read the following:

22. 110 25. 139 28. 205 31. 480
23. 125 26. 203 29. 250 32. 308
24. 108 27. 230 30. 319 33. 500

LESSON XII.

1. How many are 1 and 9?   11 and 9?   21 and 9?
31 and 9?   41 and 9?   51 and 9?   61 and 9?   71
and 9?   81 and 9?   91 and 9?

2. 2 and 9?   12 and 9?   22 and 9?   32 and 9?
42 and 9?   52 and 9?   62 and 9?   72 and 9? etc.

3. 3 and 9?   13 and 9?   23 and 9?   33 and 9? 43?
and 9? etc.

4. 4 and 9?   14 and 9?   24 and 9?   34 and 9? etc.

5. 5 and 9?   15 and 9?   25 and 9?   35 and 9? etc.

6. 6 and 9?   16 and 9?   26 and 9?   36 and 9? etc.

7. 7 and 9?   17 and 9?   27 and 9?   37 and 9? etc.

8. 8 and 9?   18 and 9?   28 and 9?   38 and 9? etc.

9. 9 and 9?   19 and 9?   29 and 9?   39 and 9? etc.

10. Count by twos till you reach 60.   Thus: two,
four, six, eight, ten, twelve, etc.

11. Count by 3s till you reach 60.

12. Count by 4s till you reach 60.

13. Count by 5s till you reach 60.

14. Count by 6s till you reach 60.

15. Count by 7s till you reach 70.

16. Count by 8s till you reach 80.

17. Count by 9s till you reach 90.

18. Count by 10s till you reach 100.

Copy and read the following :

19.	506	22.	883	25.	1,007	28.	32,368
20.	610	23.	915	26.	2,100	29.	24,345
21.	740	24.	999	27.	3,075	30.	568,073

(For explanation of terms in Addition, see p. 48.)

SUBTRACTION TABLES.

LESSON I.

To TEACHERS.—The object of this Lesson is to familiarize the learner with the natural decrease of the *first ten* numbers.

1 from	1	leaves	0.	1 from	6	leaves	5.
1 "	2	"	1.	1 "	7	"	6.
1 "	3	"	2.	1 "	8	"	7.
1 "	4	"	3.	1 "	9	"	8.
1 "	5	"	4.	1 "	10	"	9.

1. Here are 4 books: if I take 1 away, how many will be left ?

SOLUTION.—1 book from 4 books leaves 3 books.

2. If George has 3 apples, and gives 1 to his sister, how many will he have left ?

3. If Henry has 5 cents, and pays 1 cent for a pencil, how many cents will he have left ?

4. James being asked how many marbles he had, replied he had 6, lacking 1 : how many had he ?

5. If you have 8 pears, and eat 1, how many will you have left ?

6. One pen from 9 pens, leaves how many ?

7. James caught 10 butterflies, and 1 of them flew away : how many did he have left ?

8. 1 from 10 leaves how many ?   1 from 9 ?   1 from 6 ?   1 from 5 ?   1 from 4 ?   1 from 3 ?

LESSON II.

To TEACHERS.—The class can readily illustrate the Subtraction Table of 2, in the following manner:

Make 2 marks   (11); cancel 2, none are left,   ʜ

"   3   "   (111); cancel 2, one is left,   ʜ, 1.

"   4   "   (1111); cancel 2, two are left.   ʜ, 11.

"   5   "   (11111); cancel 2, three are left, ʜ, 111.

Let the class continue the illustration, and write out the Table, as below.

2 from	2	leaves	O.	2 from	7	leaves 5.
2 "	3	"	1.	2 "	8	" 6.
2 "	4	"	2.	2 "	9	" 7.
2 "	5	"	3.	2 "	10	" 8.
2 "	6	"	4.	2 "	11	" 9.

1. William had 5 oranges, and gave 2 to his sister: how many did he then have?

SOLUTION.—2 oranges from 5 oranges leave 3 oranges.

2. If you pay 3 cents for a sponge, and sell it for 2 cents, how much will you lose?

3. Two and what number are 7 ?   2 from 7 ?

4. Eight less 2 are how many ?   2 and 6 ?

5. What number is 2 less than 10? 2 more than 10?

6. Nine peaches less 2 peaches are how many ?

7. Two and what number make 11 ?

8. What is taking one number from another called?
Subtraction.

9. What is the result, or number obtained, called ?
The Difference or Remainder.

10. When we say 2 from 5 leaves 3, which is the remainder?

LESSON III.

3	from	3	leaves	0.	3 from	8 leaves	5.
3	"	4	"	1.	3 "	9 "	6.
3	"	5	"	2.	3 "	10 "	7.
3	"	6	"	3.	3 "	11 "	8.
3	"	7	"	4.	3 "	12 "	9.

1. John started with 5 pencils, but on his way to school lost 2 of them: how many did he then have?

2. If you have 6 cents, and spend 3 cents, how many cents will you have left?   3 plus 6?

3. How many are 8 less 3?   8 plus 3?

4. Show each with your fingers.

5. How many more than 3 is 5?

6. Show each with your fingers.

7. If you obtain 3 credits in the morning, how many must you get in the afternoon to make 10?

8. Sarah has 9 dollars, and her sister has only 3: how many more has one than the other?

9. Susan having 9 cents, gave 3 of them to a poor child: how many cents did she have then?

10. Twelve marbles less 3 marbles are how many?

11. Write in figures, One hundred and five, One hundred and ten, One hundred and thirty, One hundred and forty-one.

Copy and subtract the following:

(12.)	(13.)	(14.)	(15.)	(16.)	(17.)	(18.)	(19.)
12	13	14	13	16	17	18	19
3	3	3	3	3	3	3	3

LESSON IV.

4 from	4	leaves	0.	4 from	9	leaves	5.
4 "	5	"	1.	4 "	10	"	6.
4 "	6	"	2.	4 "	11	"	7.
4 "	7	"	3.	4 "	12	"	8.
4 "	8	"	4.	4 "	13	"	9.

1. If you have 7 plums, and give away 4 of them, how many will remain? Show it.

2. If 4 pinks are taken from 6 pinks, how many will remain? Show it.

3. If Henry has 8 pin-wheels, and sells 4, how many will he then have? Show it.

4. There are 10 balls on a wire of the numeral frame: if I move across 4, how many will remain?

5. If there are 12 pupils in a class, and all are perfect but 4, how many are perfect?

6. Charles had 9 chickens, but a hawk killed 4: how many has he left?

7. How many are 7 pens less 4 pens?

8. How many are 10 boys less 4 boys?

9. Amelia is 13 years old, and her sister is 4 years younger: how old is her sister?

10. Copy and read the following: 147, 175, 183, 196, 101, 105, 210, 213, 248, 260.

Copy and subtract the following:

(11.)	(12.)	(13.)	(14.)	(15.)	(16.)	(17.)	(18.)
12	13	14	15	16	17	18	19
4	4	4	4	4	4	4	4

LESSON V.

5 **from** 5 **leaves** 0.			5 **from** 10 **leaves** 5.		
5 " 6 " 1.			5 " 11 " 6.		
5 " 7 " 2.			5 " 12 " 7.		
5 " 8 " 3.			5 " 13 " 8.		
5 " 9 " 4.			5 " 14 " 9.		

1. Homer had 8 rabbits, and has sold 5 : how many has he now remaining?

2. If you have 5 pins, how many more will make 10?

3. Five and what number make 7? Show it.

4. Ten less 5 are how many? 10 plus 5?

5. George caught 11 fish, which was 5 more than his brother caught: how many did his brother catch?

6. How many are 11 less 5? 12 less 5?

7. What number taken from 9 leaves 5?

8. How many are 13 chestnuts less 5 chestnuts?

9. The price of a slate is 12 cents, and an inkstand 5 cents: what is the difference in their prices?

10. If you take 5 pears from a basket of 13 pears, how many will remain?

11. On a tree there were 11 pigeons, and a hunter shot 5 of them: how many were left?

12. Express in figures, Two hundred forty-five, Two hundred sixty, One hundred eighty-three.

Copy and subtract the following:

(13.)	(14.)	(15.)	(16.)	(17.)	(18.)	(19.)	(20.)
12	13	14	15	16	17	18	19
5	5	5	5	5	5	5	5

LESSON VI.

6 from	6 leaves	0.	6 from	11 leaves	5.
6 "	7 "	1.	6 "	12 "	6.
6 "	8 "	2.	6 "	13 "	7.
6 "	9 "	3.	6 "	14 "	8.
6 "	10 "	4.	6 "	15 "	9.

1. A party of 12 boys were skating, and 6 of them broke through the ice: how many escaped?

2. A waiter dropped a pile of 9 plates, and broke all but 6: how many did he break?

3. If you pay 8 cents for a slate and sell it for 6 cents, how much will you lose? 6 and 2?

4. How many are 10 quarts less 6 quarts?

5. Nine bananas less 6 bananas are how many?

6. Eleven birds less 6 birds are how many?

7. The price of a hat is 6 dollars, and a vest 11 dollars: what is the difference in their prices?

8. Helen is 13 years old, and is 6 years older than her brother: how old is her brother?

9. A certain class contained 15 pupils, 6 of whom were girls: how many boys were there?

10. George is now 15 years old: how old was he 6 years ago? 9 and 6 are how many?

11. Copy and read: 175, 290, 105, 210, 245, 209. 229.

Copy and subtract the following:

(12.)	(13.)	(14.)	(15.)	(16.)	(17.)	(18.)	(19.)
12	13	14	15	16	17	18	19
6	6	6	6	6	6	6	6

LESSON VII.

7	from	7	leaves	0.			
7	"	8	"	1.			
7	"	9	"	2.			
7	"	10	"	3.			
7	"	11	"	4.			

7	from	12	leaves	5.
7	"	13	"	6.
7	"	14	"	7.
7	"	15	"	8.
7	"	16	"	9.

1. Edward bought a slate for 10 cents, and paid 7 cents down: how much does he owe for it?

2. A lad having 12 eggs took them to market, and on his way broke all but 7: how many did he break?

3. How many are 11 less 7? 12 plus 7?

4. If from a dish of 13 oranges, 7 are taken out, how many will remain? 7 and 6 are how many?

5. Six from 11 leaves how many? 6 from 13? 7 from 13? 7 from 10? 7 from 11?

6. Charles had 15 young doves, and a cat killed 7 of them: how many did he then have? 7 plus 8?

7. During the last 14 school days Julia has been tardy 7 days: how many days has she been punctual?

8. Twelve is how many more than 7? What number must be added to 7 to make 12?

9. If you have 16 cents and spend 7 of them for a lunch, how many will remain?

10. Seven tops from 16 tops leave how many?

11. Express in figures, Five hundred, One thousand six hundred, Four hundred thousand.

Copy and subtract the following:

(12.)	(13.)	(14.)	(15.)	(16.)	(17.)	(18.)	(19.)
13	12	16	14	15	18	19	17
7	7	7	7	7	7	7	7

LESSON VIII.

8 from	8 leaves	0.	8 from	13 leaves	5.
8 "	9 "	1.	8 "	14 "	6.
8 "	10 "	2.	8 "	15 "	7.
8 "	11 "	3.	8 "	16 "	8.
8 "	12 "	4.	8 "	17 "	9.

1. James picked 13 quarts of chestnuts, and John 8 quarts: how many more did one pick than the other? How many quarts did both pick?

2. A father gave each of his two sons a garden spot; the elder sold his vegetables for 12 dollars, the younger sold his for 8 dollars: what was the difference in the sums each received?

3. If the older of two sisters is 16 years, and the younger 8 years, what is the difference in their ages?

4. Fourteen peaches less 3 peaches are how many?

5. Eight from 13 leaves how many? 8 plus 5?

6. A market woman bought 15 turkeys and sold 8 of them: how many had she on hand?

7. If you walk 8 miles in the morning, how far must you walk in the afternoon to make 15 miles?

8. Eight from 14, how many? 6 from 14?

9. George saved 8 yards of his kite line, which was 17 yards long: how many yards did he lose?

10. Copy and read the following: 1510, 2620, 3781, 6659, 1870, 10,807, 25,977, 85,999, 100,000.

Copy and subtract the following:

(11.)	(12.)	(13.)	(14.)	(15.)	(16.)	(17.)	(18.)
12	13	15	16	14	17	18	19
8	8	8	8	8	8	8	8

LESSON IX.

9 from	9 leaves	0.	9 from	14 leaves	5.
9 "	10 "	1.	9 "	15 "	6.
9 "	11 "	2.	9 "	16 "	7.
9 "	12 "	3.	9 "	17 "	8.
9 "	13 "	4.	9 "	18 "	9.

1. If you have 12 peanuts, how many can you eat and have 9 left?

2. Fourteen pencils less 9 pencils are how many? 14 slates less 8, are how many? 8 plus 6?

3. If you have 15 cents, how many can you spend and have 9 left?

4. William had 16 cherries upon his tree, and the birds took all but 9: how many did they take?

5. Sarah bought 15 pink roots, 9 of which died: how many lived?

6. A farmer had 14 cows in a pasture, 9 of which broke out: how many remained in the pasture?

7. What is the difference between 9 and 13? How many must be added to 9 to make 13?

8. James saw 18 butterflies upon a bed of flowers, and caught 9 of them: how many flew away?

9. There are 15 sheep in a pasture: 9 of which are white and the rest black: how many were black?

10. Express the following in figures: Two thousand and ten, Forty-five thousand six hundred.

Copy and subtract the following:

(11.)	(12.)	(13.)	(14.)	(15.)	(16.)	(17.)	(18.)
12	13	15	16	14	17	18	19
9	9	9	9	9	9	9	9

LESSON X.

Oral Drill.

1. Ten from 11 leaves how many? 10 from 21?
10 from 31? 10 from 41? 10 from 51? 10 from 61?
10 from 71? 10 from 81? 10 from 91?

2. Ten from 12 leaves how many? 10 from 22?
10 from 32? 10 from 42? 10 from 52? 10 from
62? 10 from 72? 10 from 82? 10 from 92?

3. How many are 13 less 10? 23 less 10? 33 less
10? 53 less 10? etc.

4. How many are 24 less 10? 44 less 10? 64 less
10? 34 less 10? 54 less 10? etc.

5. How many are 15 less 10? 45 less 10? 25 less
10? 55 less 10? 35 less 10? etc.

6. How many are 16 less 10? 36 less 10? 56 less
10? 46 less 10? 66 less 10? etc.

7. How many are 17 less 10? 37 less 10? 27 less
10? 47 less 10? 67 less 10? etc.

8. How many are 18 less 10? 38 less 10? 28 less
10? 48 less 10? 68 less 10? etc.

9. How many are 19 less 10? 39 less 10? 59 less
10? 29 less 10? 49 less 10? etc.

10. How many are 20 less 2? 30 less 2? 40 less
2? 50 less 2? 60 less 2? 70 less 2? etc.

11. How many are 10 less 3? 30 less 3? 20 less 3?
40 less 3? 60 less 3? 50 less 3? etc.

12. How many are 10 less 4? 30 less 4? 20 less 4?
50 less 4? 40 less 4? 60 less 4? etc.

(For explanation of terms in Subtraction, see p. 35.)

LESSON XI.

1. How many are 10 less 7 ?   30 less 7 ?   20 less 7 ?
40 less 7 ?   60 less 7 ?   50 less 7 ? etc.

2. How many are 10 less 8 ?   30 less 8 ?   20 less 8 ?
60 less 8 ?   50 less 8 ?   70 less 8 ? etc.

3. How many are 10 less 9 ?   30 less 9 ?   20 less 9 ?
40 less 9 ?   70 less 9 ?   50 less 9 ? etc.

4. 12 less 3 are how many ?   22 less 3 ?   32 less 3 ?
42 less 3 ?   52 less 3 ?   62 less 3 ? etc.

5. 13 less 5 are how many ?   23 less 5 ?   33 less 5 ?
43 less 5 ?   53 less 5 ?   63 less 5 ? etc.

6. 12 less 6 are how many?   22 less 6 ?   32 less 6 ?
52 less 6 ?   42 less 6 ?   62 less 6 ? etc.

7. 14 less 7 are how many ?   24 less 7 ?   44 less 7 ?
34 less 7 ?   54 less 7 ?   64 less 7 ? etc.

8. 14 less 5 are how many ?   24 less 5 ?   44 less 5 ?
64 less 5 ?   34 less 5 ?   54 less 5 ? etc.

9. 16 less 8 are how many ?   26 less 8 ?   46 less 8?
36 less 8 ?   66 less 8 ?   56 less 8 ? etc.

10. 15 less 9 are how many ?   25 less 9 ?   45 less 9 ?
35 less 9 ?   55 less 9 ?   75 less 9 ? etc.

11. 16 less 7 are how many?   26 less 7 ?   46 less 7 ?
36 less 7 ?   56 less 7 ?   76 less 7 ? etc.

12. 15 less 8 are how many ?   25 less 8 ?   45 less 8?
35 less 8 ?   55 less 8 ?   65 less 8 ? etc.

13. 23 less 6 are how many?   33 less 6 ?   53 less 6 ?
73 less 6 ?   63 less 6 ?   83 less 6 ? etc.

14. 17 less 8 are how many?   27 less 8 ?   47 less 8 ?
37 less 8 ?   57 less 8 ?   87 less 8 ? etc.

NOTATION.

LESSON I.

1. What is a single thing called?
A *unit or one.*

  1. What are one and one more called? "Two."

  2. Two and one more? "Three."

  3. Three and one more? "Four."

2. What are the terms one, two, three, four, &c.?
The *names of numbers.*

3. What is number?
Number is a unit, or a collection of units.

4. How are numbers generally expressed?
By *Figures, or Letters.*

5. What is expressing numbers by figures called, and why?
The *Arabic Notation.* It is so called, because it was introduced into Europe from Arabia.

6. How many figures does it employ?
The *ten* following:

1,	2,	3,	4,	5,	6,	7,	8,	9,	0.
one,	two,	three,	four,	five,	six,	seven,	eight,	nine,	naught.

7. What are the first nine called, and why?
They are called *significant figures,* because each always expresses a *number.*

They are also called *digits,* from *digitus,* a finger; because the ancients used to reckon on their fingers.

8. What is the last called, and why?
Naught, because when standing alone it has *no* value. It is also called *zero,* or *cipher.*

9. How is each of the first nine numbers expressed?
By a *single figure.*

LESSON II.

10. What are the first nine numbers called?

Units of the *first order,* or simply *units.*

11. What is the greatest number expressed by one figure ?

Nine.

12. How are numbers larger than nine expressed?

By forming *other orders of units,* called *tens, hundreds, thousands,* &c.

13. How is ten expressed?

By writing 1 in the second place, with a cipher on the right; as, 10.

14. What are figures standing in the second place, called?

Tens, or *units* of the *second order.*

15. What is the greatest number expressed by two figures ?

Ninety-nine.

16. How is a hundred expressed?

By writing 1 in the *third* place, with two ciphers on the right; as, 100.

17. What are figures standing in the third place, called ?

Hundreds, or *units* of the *third order.*

18. What are the orders higher than hundreds, called?

Thousands, tens of thousands, hundreds of thousands, millions, &c.

1. How many *simple units* make one ten ? " Ten."
2. How many *tens* make one hundred ? " Ten."
3. How many *hundreds* make one thousand? "Ten."

19. What is true as to the increase of the orders of units ?

They *increase* from right to left by the *scale of ten.* That is, ten of any *lower order* make one of the *next higher* order.

20. What places do the different orders occupy?

Simple units occupy the *right hand* place.

Tens, the second place;

Hundreds, the third place;

Thousands, the fourth place;

Tens of thousands, the fifth place;

Hundreds of thousands, the sixth place;

Millions, the seventh place, etc.

21. What is the effect of moving a figure one place from right to left, or from left to right.

Its value is *increased ten times* for every place it is moved from right to left; and is *ten times less* for every place it is moved from left to right.

22. What is the rule for expressing numbers by figures?

Begin at the left, and write the figures of the given orders in their places toward the right.

If any intermediate orders are omitted, supply their places with ciphers.

Write in figures the following numbers:

1. Thirty-seven.
2. Sixty-nine.
3. Seventy-three.
4. Eighty-five.
5. One hundred two.
6. One hundred ten.
7. One hundred five.
8. One hundred seventeen.
9. Two hundred.
10. Four hundred ten.
11. Six hundred forty-seven.
12. Eight hundred seventy-four.
13. Fifteen thousand and forty-five.
14. Forty-one thousand and ninety-five.
15. One hundred thousand and five hundred.
16. Six hundred fifty-one thousand seven hundred.
17. Eight hundred forty thousand two hundred ten.
18. One million two hundred ten thousand.

LESSON III.

1. When we designate objects as the *first, second, third, fourth*, &c., what are these terms called?
Ordinal Numbers.

2. What is the finger next to the thumb called?
" 'The first finger." The next? " The second finger."

3. The next? The next?

4. Beginning at the foot of a ladder, what is the lowest round called? The next? The next? The next? The next? What is the top one in this ladder?

· 5. Beginning at this end of the class, name the first pupil. The second. The third. The fourth. The fifth, and so on to the last.

6. With what regular number does third correspond? Fifth? Seventh?

7. How many tens in twenty? In forty? In fifty? In seventy? In sixty? In eighty? In ninety? In :. hundred?

Copy and read the following numbers.

(8.)	(9.)	(10.)	(11.)	(12.)
534	852	1264	7806	9720
638	548	3076	8520	9608
437	659	4275	9067	9999
739	947	8569	8703	. 10000

NUMERATION.

LESSON IV.

23. What is the method of reading numbers expressed by figures, called?

Numeration.

24. Beginning with units, recite the table.

Hundreds of Millions.	Tens of Millions.	*Millions.*	Hundreds of Thousands.	Tens of Thousands.	*Thousands.*	Hundreds.	Tens.	*Units.*
7	2	3,	8	7	4,	3	6	7.

Period III.	Period II.	Period

25. How read numbers expressed by figures?

Divide them into periods of three figures each, counting from the right.

Beginning at the left hand, read the periods in succession, and add the name to each, except the last.

Copy and read the following:

1. 107.	8. 20354.	15. 230684.
2. 110.	9. 23200.	16. 753007.
3. 234.	10. 43076.	17. 1000000.
4. 506.	11. 50643.	18. 5235640.
5. 730.	12. 62640.	19. 18642065.
6. 809.	13. 84063.	20. 81000000.
7. 943.	14. 97810.	21. 463250648.

☞ If this and the next three lessons are deemed too difficult for beginners, they may be omitted till review.

ROMAN NOTATION.

LESSON V.

26. By what other method are numbers expressed?

By the *following letters*, viz.: I, V, X, L, C, D, M.

27. What does each of these letters denote?

The letter I, denotes one; V, five; X, ten; L, fifty; C, one hundred; D, five hundred; M, one thousand.

28. How are other numbers expressed by these letters?

By *repeating* and combining them, as in the following

TABLE.

I	denotes one.	XXIV	denotes twenty-four, &c.
II	" two.	XXX	" thirty.
III	" three.	XXXI	" thirty one.
IV	" four.	XXXII	" thirty-two, &c.
V	" five.	XL	" forty.
VI	" six.	XLI	" forty-one, &c.
VII	" seven.	L	" fifty.
VIII	" eight.	LX	" sixty.
IX	" nine.	LXX	" seventy.
X	" ten.	LXXX	" eighty.
XI	" eleven.	XC	" ninety.
XII	" twelve.	C	" one hundred.
XIII	" thirteen.	CX	" one hundred ten.
XIV	" fourteen.	CC	" two hundred.
XV	" fifteen.	CCC	" three hundred.
XVI	" sixteen.	CCCC	" four hundred.
XVII	" seventeen.	D	" five hundred.
XVIII	" eighteen.	DC	" six hundred.
XIX	" nineteen.	DCC	" seven hundred.
XX	" twenty.	DCCC	" eight hundred.
XXI	" twenty-one.	DCCCC	" nine hundred.
XXII	" twenty-two.	M	" one thousand.
XXIII	" twenty-three.	MD	" one thou. five hund.

MDCCCLXXV, one thousand eight hundred and seventy-five.

LESSON VI.

29. When a letter is repeated, what is the effect?

Its value is repeated. Thus, I denotes one; II, two; X, ten; XX, twenty, &c.

30. If a letter is placed *before* one of greater value, what is the effect?

The *value of the less* is *taken* from the greater.

31. If a letter is placed *after* one of greater value, what?

The *value of the less* is *added* to the greater.

1. What does X denote? What IX? What XI?

2. What does V denote? What IV? What VI?

32. How express the numbers from 10 to 20 by letters?

By *adding the letters* of the first decade to X; as X, XI, XII, &c.

Express the following numbers by letters: 7, 11, 14, 19, 29, 39, 41, 40, 60, 70, 89, 91, 101, 550, 670, 1010, 1500, 1875.

Copy and read the following numbers.

1. IV.	7. XXXIX.	13. CIV.
2. VI.	8. XL.	14. CCC.
3. VII.	9. LX.	15. DCX.
4. XIV.	10. LIX.	16. MDL.
5. XVI.	11. LXXIV.	17. MDCC.
6. XXIV.	12. LXXXVIII.	18. MDCCCV.

33. In what is the Roman Notation chiefly used?

In expressing the sections, chapters. lessons, &c. into which books are divided, and in marking the hours on the faces of clocks and watches.

LESSON VII.

To Tell the Time of Day by the Clock.

34. How is the face of a clock divided?

Ans. Into *twelve equal* parts called hours, marked by the letters, I, II, III, etc.

35. What is the object of the two pointers, or hands.

Ans. The *short* hand tells the hours, and is called the *hour hand;* the *long* one tells the minutes, and is called the *minute hand.*

1. When both hands are at XII, what time is it?

Ans. It is twelve o'clock.

NOTE.—The teacher will explain that when the minute hand reaches II, it is ten minutes past one o'clock; when it reaches III, it is fifteen minutes past one, and while the hour hand passes from XII to I, the minute hand moves entirely around the face, and points at XII, &c.

2. When the minute hand is at VI and the hour hand half way between II and III, what is the time?

Ans. It is half past two o'clock

3. When the minute hand points at IX, what is the time?

Ans. It lacks fifteen minutes of three o'clock.

4. Where must the hands be, to denote ten minutes past five?

5. Where must they be, to denote half past five?

6. Where must they be, to denote 20 minutes past eight?

7. When the hour hand is past III, and the minute hand is at V, what time is it?

8. When the hour hand is near VI and the minute hand at IX, what?

ADDITION.

LESSON I

1. What is Addition?

Addition is uniting two or more numbers in one.

2. What is the number obtained by addition called?

The *Sum* or *Amount.*

3. How is Addition denoted?

By a *perpendicular cross* called *plus* (+), placed between the numbers to be added. Thus, the expression $4+3$ shows that 4 and 3 are to be added together, and is read, "4 plus 3," "4 and 3," or "4 added to 3."

NOTE.—The term *plus* signifies *more* or *added to.*

4. How is the equality between numbers denoted?

By *two short parallel lines,* called the *sign of equality* (=). The expression $4 + 3 = 7$, shows that 4 increased by 3 equals 7, and is read, "4 plus 3 equal 7," or the sum of "$4 + 3$ equals 7."

Copy and read the following expressions:

1. $5 + 3 + 2 + 0 + 8 = 10 + 3 + 5.$

2. $6 + 8 + 0 + 9 = 2 + 10 + 3 + 8.$

3. One pupil gave her teacher 7 peaches, another 8, and another 6: how many peaches did all give her?

SOLUTION.—7 peaches and 8 peaches are 15 peaches, and 6 are 21 peaches: therefore all gave her 21 peaches.

4. If John picks 5 roses from one bush, 9 from another, and 7 from another, how many will he pick?

5. How many are 6 quarts, 8 quarts, and 7 quarts?

6. How many are 18 yards and 9 yards and 4 yards?

LESSON II.

To add single columns when two or more numbers, coming together, make 10.

1. Find the amount of 5, 6, 3, 7, 8, 4, 6, 9.

ANALYSIS.—Write the numbers one under another, in a perpendicular column, and draw a line under it.

Beginning at the bottom and omitting the names of the numbers, proceed thus: Nine, nineteen (adding 10 for 6 and 4), twenty-seven, thirty-seven (adding 10 for 7 and 3), forty-three, forty-eight. The amount is 48.

NOTES.—1. When two or more numbers together make *ten*, instead of adding these numbers separately, it is *better* to add 10 at once.

2. In all operations both mental and slate, the pupil should add each number as a *whole*, and not by *single units*, or by counting his fingers. Counters should be used no longer than necessary to illustrate the different combinations.

3. Great care should be taken to see that the figures are written with *neatness* and *symmetry*, and in *perpendicular* columns.

Operation.

```
 5
 6
 3
 7
 8
 4
 6
 9
 —
48  Ans.
```

Copy and add the following in like manner:

(2.)	(3.)	(4.)	(5.)	(6.)	(7.)	(8.)	(9.)
8	4	3	9	4	7	6	8
2	6	2	4	7	3	9	7
7	5	6	3	2	6	8	9
2	2	4	2	3	4	7	6
3	6	3	8	5	5	2	8
7	5	7	6	8	2	5	7
6	5	5	3	2	3	3	8
6	3	6	2	3	8	2	7
4	7	4	5	6	7	7	8
7	8	3	4	5	6.	8	9

LESSON III.

Mental Exercises.

1. A teacher received 6 apples from one of her pupils, 7 from another, and 8 from another: how many apples did she receive from all?

2. If you pay 10 cents for a slate, 6 cents for an inkstand, and 3 cents for a pencil, how much will you pay for all?

3. If an orange costs 6 cents, a pear 5 cents, and a lemon 4 cents, what will they all cost?

4. How many are 7 brooms, 3 brooms, and 8 brooms?

5. How many are 8 days, 7 days, and 6 days?

6. How many are 7, and 6, and 8?

7. If Harry receives 6 credits a day for 3 days, how many marks will he have?

8. How many are 9 dollars, 6 dollars and 7 dollars?

9. How many are 12, and 8, and 9?

Copy and add the following:

(1.)	(2.)	(3.)	(4.)	(5.)	(6.)	(7.)	(8.)
6	7	8	9	10	8	6	7
6	7	8	9	10	9	7	9
6	7	8	9	10	6	8	8
6	7	8	9	10	5	6	6
6	7	8	9	10	8	7	7
6	7	8	9	10-	9	8	8
6	7	8	9	10	7	6	6
6	7	8	9	10	3	7	7
6	7	8	9	10	8	8	8
6	7	8	9	10	9	9	6

LESSON · IV.

To add numbers consisting of two or more Columns, when the sum of each column is less than 10.

1. What is the sum of 223, 342, and 132 ?

ANALYSIS.—Write the numbers one under another, units under units, tens under tens, etc., and, beginning at the right, add thus : 2 units and 2 units are 4 units, and 3 are 7 units. Set the 7 under the units' place, *because it is units.* Next, 3 tens and 4 tens are 7 tens, and 2 are 9 tens. Set the 9 in tens' place, *because it is tens.*

Operation.
223
342
132
———
697 *Ans.*

Finally, 1 hundred and 3 hundreds are 4 hundreds, and 2 are 6 hundreds. Set the 6 in hundreds' place, *because it is hundreds.*   *Ans.* 697.

(2.)	(3.)	(4.)	(5.)	(6.)	(7.)
31	23	24	231	312	413
24	32	30	324	431	150
11	22	23	132	205	326

(2.)	(3.)	(4.)	(5.)	(6.)	(7.)
25	23	30	231	324	403
30	42	43	123	230	232
24	34	25	404	305	354

2. A farmer has two flocks of sheep, one containing 342, the other 227 : how many sheep has he ?

3. Write in columns and find the sum of 313 dollars, 142 dollars, and 432 dollars.

4. James has three books, one containing 212 pages, another 320 pages, and another 456 pages : how many pages do all contain ?

LESSON V.

To add numbers consisting of two or more Columns, when the sum of a column is 10, or more.

1. What is the sum of 234 dollars, 525 dollars, and 443 dollars?

ANALYSIS.—Write the numbers one under another, the units under units, etc., and, beginning
at the right, add as before. Thus, 3 units and 5
units are 8 units, and 4 are 12 units, equal to 1 ten
and 2 units. Set the 2 in units' place, because it
is units, and add the 1 ten to the next column, because it is the *same order* as that column.

Operation.
234 d.
525 d.
443 d.
———
Ans. 1202 d.

Next, 1 ten and 4 tens are 5 tens, and 2 are 7 tens, and 3 are 10
tens, equal to 1 hundred and 0 tens. Set the 0, or unit's figure,
under the column added, *because there are no tens*, and add the
1 hundred to the next column, because it is the *same order* as
that column. Adding the 1 hundred to the next column, the
sum is 12 hundreds, and this being the last column, we set
down the *whole sum*.

NOTE.—As soon as the learner becomes familiar with adding
numbers which have two or more columns, he should omit the
name of the order, etc., and pronounce the results only, as in
adding single columns. (P. 38, Ex. 1.)

2. What two principles are necessary to be observed
in addition?

1st. The numbers must be *Like Numbers.*

2d. *Units* of the *same order* must be added,
each to each.

3. What are like numbers?

Like Numbers are those which express *units*
of the same *kind;* as, 4 pears and 3 pears; 5 and 8, etc.

4. What are unlike numbers?

Unlike Numbers are those which express units
of different kinds; as, 4 dollars and 3 yards.

LESSON VI. .

Review of Principles.

5. How do you write numbers to be added?

Write one under another, units under units, etc.

6. When begin to add, and how proceed?

Begin at the right, and add each column separately.

7. When the sum of a column is less than 10, what is done with it; and why?

Set it under the column added, because it is the same order as that column.

8. When the sum of a column *exceeds* 9, what do you do with it?

Write the units' figure under the column added, and add the tens to the next higher order.

9. What do you do with the last column?

Set down the whole sum.

10. How is Addition proved?

Begin at the top, and add each column downward. If the two results agree, the work is right.

Examples for Practice.

(1.)	(2.)	(3.)	(4.)	(5.)
233	234	395	382	504
165	364	265	237	160
486	246	486	68	439
283	547	257	385	758

6. Find the sum of 305 yds., 28 yds., and 420 yds.

7. Find the sum of 325 dols., 83 dols., and 7 dols.

8. What is the sum of $436 + 48 + 136 + 20$?

9. What is the sum of $3450 + 243 + 1789 + 46$?

| (10.) | (11.) | (12.) | (13.) | (14.) |
Dollars.	Feet.	Days.	Pounds.	Gallons.
3465	4273	612	7260	8725
802	6250	7309	39	430
5060	367	527	547	57
2432	5046	6305	9084	5367

15. A man picked 875 oranges from one tree, 739 from another, and 1237 from another: how many did he pick from all?

16. One school has 475 pupils, another 630, and another 568: how many pupils have all?

(17.)	(18.)	(19.)	(20.)	(21.)
4358	3460	4504	6720	8354
754	58	75	4857	2075
5243	8539	7322	5081	4630

22. Find the sum of 275 pounds + 468 pounds + 723 pounds.

23. Find the sum of 463 yds. + 568 yds. + 837 yds.

24. Find the sum of 563 gal. + 645 gal. + 750 gal.

(25.)	(26.)	(27.)		(28.)		(29.)		(30.)	
30	73	3	25	4	84	35	56	67	45
48	30	4	76	5	40	93	27	90	76
53	38	8	25	3	33	82	82	38	60
37	45	7	30	2	22	80	75	54	93
72	72	6	34	3	58	64	40	73	40
45	64	4	25	7	34	95	67	85	63
38	93	7	46	2	76	38	53	64	88
53	78	5	63	8	43	47	48	90	70

SUBTRACTION.

LESSON I.

1. What is Subtraction?

Subtraction is *taking one number from an other.*

2. What is the number to be subtracted called?

The *Subtrahend.*

3. The number from which the subtraction is made?

The *Minuend.*

4. What is the number obtained by Subtraction called?

The *Difference,* or *remainder.*

1. When we say, 3 from 8 leaves 5, which is the minuend? The subtrahend? The remainder?

2. When it is said that 6 taken from 14 leaves 8, what is the 6 called? The 14? The 8?

5. How is Subtraction denoted?

By a *short horizontal line,* (−) called *minus.* When placed between two numbers, this sign shows that the number *after* it is to be taken from the one *before* it. Thus, 5 − 3 shows that 3 is to be taken from 5, and is read " 5 minus 3."

NOTE.—The term *minus* signifies *less.*

Copy and read the following expressions:

1. $8 - 3 = 10 - 5.$ 4. $119 - 5 = 118 - 4.$
2. $23 - 5 = 16 + 2.$ 5. $135 + 8 = 150 - 7.$
3. $87 + 4 = 98 - 7$ 6. $250 + 7 = 277 - 20.$

7. 19 dollars − 7 dollars = how many dollars?
8. 15 bushels − 8 bushels = how many bushels?
9. $23 - 8 =$ how many? $27 - 10 =$ how many?

LESSON II.

When each Figure in the Lower Number is Less than the one above it.

1. Find the difference between 746 and 214.

ANALYSIS.—Write the *less* number under the 　*Operation.*
greater, units under *units, tens* under *tens*, etc.　　746
Beginning at the right, proceed thus: 4 units　　　214
from 6 units leave 2 units. Set the 2 in units'　　——
place, under the figure subtracted, because it is 　*Ans.* 532
units. Next, 1 ten from 4 tens leaves 3 tens. Set
the 3 in tens' place, under the figure subtracted, because it is
tens. Finally, 2 hundreds from 7 hundreds leave 5 hundreds.
Set the 5 under the hundreds' column, because it is *hundreds.*

Solve the following in a similar manner:

(2.)	(3.)	(4.)	(5.)	(6.)
435	546	615	768	879
312	221	314	544	636

(7.)	(8.)	(9.)	(10.)
575 dols.	465 yards.	675 days.	743 acres.
243 dols.	234 yards.	372 days.	423 acres.

11. A farmer having 456 sheep, sold 230 of them: how many did he have left?

12. If a man's income is 685 dollars a year, and his expenses are 360 dollars, how much can he lay up?

13. What is the difference between 570 and 340?

14. What is the difference between 700 and 300?

15. What is the difference between six hundred forty-five and two hundred twenty-three?

(16.)	(17.)	(18.)	(19.)
456 dols.	564 sheep.	678 feet.	784 pounds.
324 dols.	234 sheep.	538 feet.	541 pounds.

(20.)	(21.)	(22.)	(23.)
5674	7360	8679	9230
2351	4230	5360	4020

Mental Exercises.

1. A tailor sold a coat for 25 dollars, and received 10 dollars down : how much is due him ?

ANALYSIS.—Ten dollars from 25 dollars leave 15 dollars.

2. If you have 17 doves, and sell 8 of them, how many will you have left ?

3. Bought a cow for 35 dollars, and sold her for 9 dollars less than cost : for how much was she sold ?

4. A market boy had 44 eggs in a basket, and letting it fall broke 12 : how many remained unbroken ?

5. William had a 25-cent piece to buy a lunch, the price of which was 15 cents : how much change should he receive ?

6. A young man is 21 years old to-day : how old was he 8 years ago ?

7. Nine and what number are 23 ? 8 and what number are 32 ?

8. 7 and what are 35 ? 6 and what are 42 ?

9. Henry has 10 dollars in the savings-bank : how many dollars more must he get to make 50 dollars ?

10. There are 25 cows in a pasture : if 9 are taken out, how many will be left ?

LESSON III.

When a Figure in the Lower Number is Larger than the one above it.

1. What is the difference between 745 and 438 ?

ANALYSIS.—Here 8 is larger than 5, and can- *Operation.* not be taken from it. How is this difficulty re- 745 moved ? There are two methods. 438

1st METHOD.—We add 10 to the 5, making 15 ; —— now 8 units from 15 units leave 7 units. We *Ans.* 307 write the 7 in units' place under the figure sub- tracted. To balance the 10 added to the upper number, we add 1 to the next higher order of the lower, which is equal to the 10 added to the upper number. Now 1 ten and 3 tens are 4 tens, and 4 tens from 4 tens leave 0 tens. Place a cipher in tens' place, because there are 0 tens. Finally, 4 hundreds from 7 hundreds leave 3 hundreds, which we write in hundreds' place. The remainder is 307.

2d METHOD.—Instead of 10 taken at random, we may take 1 ten from the 4 tens in the upper number, and add it to the 5 units, making 15. Now 8 from 15 leaves 7, which we set under the figure subtracted. Since we took 1 ten from the 4 tens, there are but 3 tens left ; and 3 tens from 3 tens leave 0 tens. Finally, 4 from 7 leaves 3. The remainder is 307, the same as before.

NOTE.—The process of adding 10 to the upper figure is commonly called " borrowing ten."

6. What two principles are necessary to be observed in subtraction ?

1st. The numbers must be *Like Numbers.*

2d. *Units of the same order* must be sub- tracted one from the other.

LESSON IV.

Review of Principles.

7. How do you write numbers for subtraction ?

Write the less number under the greater, units under units, tens under tens, etc.

8. Where do you begin to subtract, and how proceed ?

Begin at the right, and subtract each figure in the lower number from the one above it, setting the remainder under the figure subtracted.

9. If a figure in the lower number is larger than the one above it, how proceed ?

Add 10 to the upper figure; then subtract, and add 1 to the next figure in the lower number.

10. How is subtraction proved ?

Add the remainder to the subtrahend; if the sum is equal to the minuend, the work is right.

Examples for Practice.

1. Find the difference between 745 and 280; and prove the operation.

(2.)	(3.)	(4.)	(5.)
234 dols.	435 barrels.	647 gallons.	730 days.
108 dols.	260 barrels.	365 gallons.	365 days.

(6.)	(7.)	(8.)	(9.)
5450	6305	5785	7346
2237	3252	3060	5037

(10.)	(11.)	12.)	(13.)
345 yards.	520 dols.	671 pounds.	784 quarts.
160 yards.	235 dols.	486 pounds.	92 quarts.

(14.)	(15.)	(16.)	(17.)
3427	6504	8050	9650
1285	4273	4370	5645

18. A farmer raised 6256 bushels of wheat, and sold 3460 : how much remained unsold?

19. The price of a house is 6475 dollars, and that of a farm 7500 dollars : what is the difference in the prices?

20. What is the difference between five thousand and twenty-five, and twenty-five hundred and five?

21. What is the difference between ten thousand and ten, and ten hundred and ten?

22. How many are 4560 dollars less 2345 dollars?

23. Bought a lot of goods for 13250 dollars, and sold them for 12500 dollars : what was the loss?

24. 63256 — 500200 ?   27. 710237 — 500420?

25. 70240 — 43210 ?   28. 806430 — 650340 ?

26. 85207 — 60340 ?   29. 900645 — 704306 ?

30. The population of New York in 1860 was 3.880,735; in 1870 it was 4,370,846: what was the increase?

31. The population of the United States in 1860 was 31,443,321; in 1870 it was 38,312,633: what was the increase?

32. A man paid $2500 dollars for a farm, and sold it for 4000 dollars: how much did he make?

LESSON V.

Oral Drill.

1. To 4 add 6; subtract 3; add 9; subtract 5; add 12; subtract 3; add 4; subtract 12; add 8; subtract 5: what is the result?

EXPLANATION.—The teacher says, "to 4 add 6," the class think 10; "subtract 3," the class think 7; "add 9," the class think 16; "subtract 5," the class think 11; "add 12," the class think 23; "subtract 3," the class think 20, etc.

2. From 8 subtract 2; add 4; subtract 5; add 7; subtract 4; add 2; subtract 5; add 4; subtract 6: result?

3. To 12 add 5; subtract 4; add 6; subtract 8; add 7; subtract 4; add 7; subtract 6; add 4: result?

4. From 13 take 4; add 8; take 5; add 7; take 6; add 8; take 8; add 7; take 5; add 6: result?

5. From 14 take 7; add 4; take 3; add 5; take 9; add 6; take 8; add 3; take 4; add 8: result?

6. To 15 add 5; subtract 6; add 4; subtract 7; add 6; add 5; subtract 6; add 7; subtract 8; add 9: result?

7. From 20 take 3; add 8; take 6; add 9; take 8; add 7; take 9; add 6; take 5; add 6: result?

8. From 34 take 3; add 6; take 10; add 2; take 7; add 5; take 4; add 9: result?

9. To 17 add 8; take 6; add 10; take 8; take 3; add 7; take 4; add 6: result?

10. From 43 take 7; add 4; add 9; take 6; take 5; add 10; add 20; take 6: result?

MULTIPLICATION.

LESSON I.

To TEACHERS.—The object of this Lesson is to *develop* the *idea* of *times*, as used in Multiplication, and lead the class to see the similarity of Multiplication to Addition.

1. Each pupil may make a star or unit mark upon his slate, as I make one upon the blackboard. ✳

2. How many times have you made one star ?
 One time.

3. Make another under the first.   ✳

4. How many times have you made 1 star now ?
 Two times.

5. Make 2 groups of 2 stars each; as, ✳ ✳, ✳ ✳.

6. How many times have you made 2 marks ?
 "2 times."

7. How many are 2 marks and 2 marks ?
 "4 marks."

8. How many are 2 times 2 marks?   "4 marks."

9. The next make 3 groups of 2 stars each; as,
 ✳ ✳,   ✳ ✳,   ✳ ✳.

10. How many times have you made 2 stars ?
 Three times.

11. How many are 3 times 2 marks ?   "6 marks."

12. Make 4 groups of 2 stars each; as,
 ✳ ✳,   ✳ ✳,   ✳ ✳,   ✳ ✳.

13. How many are 4 times 2 stars ?   "8 stars."

LESSON II.

2 times	1	are	2.	2 times	7	are	14.
2 "	2	"	4.	2 "	8	"	16.
2 "	3	"	6.	2 "	9	"	18.
2 "	4	"	8.	2 "	10	"	20.
2 "	5	"	10.	2 "	11	"	22.
2 "	6	"	12.	2 "	12	"	24.

1. If 1 lemon costs 6 cents, what will 6 lemons cost?

ILLUSTRATION.—Let the pupil place a group of 6 stars upon the blackboard, representing 6 cents, the price of 1 lemon; then will 2 groups of 6 stars each, represent the cost of 2 lemons; as,

✳ ✳ ✳ ✳ ✳ ✳ ;   ✳ ✳ ✳ ✳ ✳ ✳ .

Counting the stars in these 2 groups together, we have 12 stars; hence, the cost of 2 lemons is 12 cents. Or, by

ANALYSIS.—If 1 lemon costs 6 cents, 2 lemons will cost 2 times 6 cents; and 2 times 6 cents are 12 cents.

2. How many are 2 times 4 cents?

3. Paid 2 dollars apiece for 2 caps: how much did both caps cost? Show it.

4. How many are 2 times 3 fingers? Show it.

5. At 6 dollars apiece, what will 2 chairs cost?

6. What cost 2 desks, at 5 dollars apiece?

7. If you ride 9 miles an hour, how far will you ride in 2 hours?

8. What cost 2 barrels of flour, at 10 dollars a barrel?

LESSON III.

3	times	1	are	3.	3	times	7	are 21.
3	"	2	"	6.	3	"	8	" 24.
3	"	3	"	9.	3	"	9	" 27.
3	"	4	"	12.	3	"	10	" 30.
3	"	5	"	15.	3	"	11	" 33.
3	"	6	"	18.	3	"	12	" 36.

1. Louisa bought 3 lead pencils at 5 cents each: how much did she pay for all?

ANALYSIS.—Since 1 pencil cost 5 cents, 3 pencils will cost 3 times 5 cents; and 3 times 5 cents are 15 cents.

2. What will 3 apples cost, at 2 cents apiece?

3. There are 3 feet in 1 yard: how many feet are there in 3 yards?

4. How many are 3 times 3 rabbits?

5. Bought 3 peaches at 4 cents each: what did they come to?

6. If a car goes 6 miles an hour, how far will it go in 3 hours?

7. What cost 3 inkstands, at 8 cents each?

8. Paid 7 dollars apiece for 3 vests: what was the cost of all?

9. Henry has 3 coops of 9 chickens each: how many chickens has he in all?

10. What is finding the amount of a number taken or added to itself a given number of times called?

Multiplication.

11. What is the result obtained by multiplication called?

LESSON IV.

4	—	1	—	4.	4	—	7	—	28.
4	"	2	"	8.	4	"	8	"	32.
4	"	3	"	12.	4	"	9	"	36.
4	"	4	"	16.	4	"	10	"	40.
4	"	5	"	20.	4	"	11	"	44.
4	"	6	"	24.	4	"	12	"	48.

1. What is the difference in the cost of 4 pens, at 3 cents apiece, and 3 lemons, at 4 cents apiece?

ILLUSTRATION.—Let the pupil place upon the board 4 groups of 3 stars each, and suppose them to represent the cost of 4 pears, at 3 cents apiece: as,

✻ ✻ ✻   ✻ ✻ ✻   ✻ ✻ ✻   ✻ ✻ ✻

Also, by 3 groups of 4 stars each, let him represent in like manner the cost of the lemons; as,

✻ ✻ ✻ ✻   ✻ ✻ ✻ ✻   ✻ ✻ ✻ ✻

Counting the stars in these rows separately, we find that each row contains 12 stars. Hence, 4 times 3 cents is the same as 3 times 4 cents; consequently there is no difference in the cost.

2. What cost 4 pair of boots, at 5 dollars a pair?

3. If you pay 6 dollars a week for board, what will it cost you to board 4 weeks?

4. Which is greater, 2 times 6, or 6 times 2? Show it.

5. What cost 4 chairs, at 7 dollars each?

6. At 12 cents each, what cost 4 slates?

Copy and multiply the following:

(7.)	8.	(9.)	(10.)
312	521	602	821
4	4	4	4

LESSON V.

5	times	1	are	5.	5	times	7	are	35.
5	"	2	"	10.	5	"	8	"	40.
5	"	3	"	15.	5	"	9	"	45.
5	"	4	"	20.	5	"	10	"	50.
5	"	5	"	25.	5	"	11	"	55.
5	"	6	"	30.	5	"	12	"	60.

1. George picked 5 quarts of blackberries, and sold them at 6 cents a quart: what did they come to?

2. In 1 gal. are 4 quarts: how many quarts in 5 gallons?

3. What cost 5 bananas, at 5 cents each?

4. George bought 6 kites, at 5 cents apiece: what did he pay for them all?

5. How many bushels of apples will 5 apple trees bear, allowing 7 bushels to a tree?

6. What cost 5 pair of boots, at 8 dollars a pair?

7. How many problems will you solve in 5 days, if you solve 10 each day?

8. What is the cost of 5 tables, at 9 dollars apiece?

9. A furrier sold 5 muffs, at 12 dollars apiece: how much did he receive for them?

10. If 11 yards of muslin will make 1 dress, how many will be required to make 5 dresses?

11. Write six thousand, four hundred and one.

Copy and multiply the following:

(12.)	(13.)	(14.)	(15.)	(16.)
410	501	610	700	811
5	5	5	5	5

LESSON VI.

6	times	1	are	6.	6	times	7 are	42.
6	"	2	"	12.	6	"	8 "	48.
6	"	3	"	18.	6	"	9 "	54.
6	"	4	"	24.	6	"	10 "	60.
6	"	5	"	30.	6	"	11 "	66.
6	"	6	"	36.	6	"	12 "	72.

1. What will 6 chairs cost, at 4 dollars apiece?

2. Sold 5 quarts of cherries, at 6 cents a quart: what did they come to?

3. If 6 yards of silk will make 1 cloak, how many yards will it take to make 6 cloaks?

4. In 1 week are 7 days: how many days in 6 weeks?

5. How many bushels are 6 times 8 bushels?

6. At 9 dollars apiece, what will 6 desks cost?

7. A party of 6 children gathered 10 quarts of nuts apiece: how many did they all gather?

8. A pleasure party hired 6 sail boats; each boat carried 11 persons: how many in the party?

9. How many pounds of honey are there in 6 boxes, if each box contains 12 pounds?

10. What will 6 penknives cost, at 12 cents each?

11. Write twelve thousand, and three hundred.

12. Write ten thousand, two hundred and six.

Copy and multiply the following:

(13.)	(14.)	(15.)	(16.)	(17.)
5111	6110	7100	8000	9010
6	6	6	6	6

LESSON VII.

7 times	1	are	7.		7 times	7	are	49.
7	"	2	"	14.	7	"	8	"	56.
7	"	3	"	21.	7	"	9	"	63.
7	"	4	"	28.	7	"	10	"	70.
7	"	5	"	35.	7	"	11	"	77.
7	"	6	"	42.	7	"	12	"	84.

1. If 1 settee holds 6 pupils, how many can sit on 7 ?
2. At 7 dollars a barrel, what cost 4 barrels of flour ?
3. What cost 7 slates, at 8 cents apiece ?
4. Homer paid 5 cents apiece for 7 oranges: how much did he pay for all?
5. Are 7 times 5 greater or less than 6 times 7 ?
6. If a stage goes 7 miles an hour, how far will it travel in 7 hours?
7. A farmer sold 7 tons of hay at 11 dollars a ton: how much did he receive for his hay ?
8. How many quarts will 7 cherry trees bear, if each tree bears 10 quarts?
9. If a clerk earns 9 dollars a week, how much will he earn in 7 weeks?
10. Are 7 times 7 more or less than 6 times 8 ?
11. At 12 cents each, what will 7 melons come to ?
12. Write forty thousand, five hundred seventy.

Copy and multiply the following:

(13.)	(14.)	(15.)	(16.)	(17.)
3101	5011	4111	8001	9100
3	7	7	7	7

LESSON VIII.

8	times	1	are	8.	8	times	7	are 56.
8	"	2	"	16.	8	"	8	" 64.
8	"	3	"	24.	8	"	9	" 72.
8	"	4	"	32.	8	"	10	" 80.
8	"	5	"	40.	8	"	11	" 88.
8	"	6	"	48.	8	"	12	" 96.

1. What cost 8 quarts of cider, at 4 cents a quart?

2. If 5 school days constitute a week, how many school days are in 8 weeks?

3. George spent 6 cents a day for his lunch: how much did he spend in 8 days?

4. Sold 7 barrels of flour, at 8 dollars a barrel: what did it come to?

5. What cost 8 yards of cloth, at 8 dollars a yard?

6. Are 8 times 5 more or less than 6 times 7?

7. In 1 mile there are 8 furlongs: how many furlongs are there in 9 miles?

8. If 8 young peach trees bear 10 peaches each, how many will all bear?

9. A teacher has 8 classes in her school, with 11 scholars in each class: how many pupils has she?

10. Which is greater, 8 times 9, or 6 times 12?

11. Write two hundred and ten thousand and seven.

Copy and multiply the following:

(12.)	(13.)	(14.)	(15.)	(16.)
5100	7010	6110	8011	9111
8	8	8	8	8

LESSON IX.

9	times	1	are	9.	9	times	7	are 63.
9	"	2	"	18.	9	"	8	" 72.
9	"	3	"	27.	9	"	9	" 81.
9	"	4	"	36.	9	"	10	" 90.
9	"	5	"	45.	9	"	11	" 99.
9	"	6	"	54.	9	"	12	" 108.

1. In 1 yard are 3 feet: how many feet in 9 yards?

2. What cost 4 tables, at 9 dollars each?

3. How far will a boat sail in 9 hours, if she sails 5 miles an hour?

4. If a hunter kills 6 pigeons at a shot, how many will he kill by 9 shots?

5. What cost 9 barrels of nuts, at 7 dols. a barrel?

6. Paid 9 cents apiece for 8 inkstands: what did they all come to?

7. What is the difference between 9 times 6 and 7 times 8?

8. At 10 cts. a quart, what cost 9 quarts of berries?

9. In 1 dollar there are 10 dimes: how many dimes are there in 9 dollars?

10. Charles obtained 9 merits, and received 12 cents for each merit: how many cents did he receive?

11. Which is greater, 9 times 8, or 7 times 9?

12. Write one hundred and six thousand.

Copy and multiply the following:

(13.)	(14.)	(15.)	(16.)
6101	5001	8011	9111
9	9	9	9

LESSON X.

10 times	1	are	10.		10 times	7	are	70.
10	"	2	"	20.	10	"	8	"	80.
10	"	3	"	30.	10	"	9	"	90.
10	"	4	"	40.	10	"	10	"	100.
10	"	5	"	50.	10	"	11	"	110.
10	"	6	"	60.	10	"	12	"	120.

1. What cost 10 tons of coal, at 6 dollars a ton?
2. At 10 cents a yard, what cost 7 yards of muslin?
3. If a man earns 10 dollars a week, how much will he earn in 8 weeks?
4. At 10 dimes each, how many dimes in 9 dollars?
5. What is the cost of 10 tables, at 10 dollars apiece?
6. At 12 dollars apiece, what cost 10 overcoats?
7. How many bushels shall I have from 10 apple trees, if each tree bears 11 bushels?
8. Write eight hundred, and multiply it by 10.

LESSON XI.

11 times	1	are	11.		11 times	7	are	77.
11	"	2	"	22.	11	"	8	"	88.
11	"	3	"	33.	11	"	9	"	99.
11	"	4	"	44.	11	"	10	"	110.
11	"	5	"	55.	11	"	11	"	121.
11	"	6	"	66.	11	"	12	"	132.

1. A farmer sent 11 chickens to market, each weighing 3 pounds: how many pounds did they all weigh?
2. What cost 11 thimbles, at 4 cents each?
3. At 6 cents each, what will 11 tops come to?

4. What cost 11 pounds of grapes, at 10 cents a pound?

5. How many apple trees are there in an orchard, which has 7 rows and 11 trees in a row?

6. If 1 melon is worth 8 peaches, how many peaches are 11 melons worth?

7. If you write 9 lines each day, how many will you write in 11 days?

8. At 12 cents apiece, what will 11 knives cost?

LESSON XII.

12 times	1 are	12.	12 times	7 are	84.
12 "	2 "	24.	12 "	8 "	96.
12 "	3 "	36.	12 "	9 "	108.
12 "	4 "	48.	12 "	10 "	120.
12 "	5 "	60.	12 "	11 "	132
12 "	6 "	72.	12 "	12 "	144.

1. William has 5 marbles, and James has 12 times as many: how many marbles has James?

2. How many roses will 4 bushes produce, if each bush has 12 roses?

3. At 6 cents a mile, what will it cost to ride 12 miles.

4. At 7 dollars a week, how much will a man spend in 12 weeks?

5. In 1 dozen there are 12 units: how many units are there in 9 dozen?

6. What cost 10 dozen eggs, at 12 cents a dozen?

7. A car has 8 wheels: how many wheels to a train of 12 cars?

LESSON XIII.

Explanation of Terms.

1. What is Multiplication?

Multiplication is finding the *amount* of a number taken or added to itself, a given number of times.

2. What is the number to be multiplied, called?

The *Multiplicand.*

3. What the number by which you multiply?

The *Multiplier;* and shows how many times the multiplicand is to be taken?

4. What is the number obtained by multiplication called?

The *Product.*

When it is said that 2 times 3 are 6, which is the multiplicand? The multiplier? The product?

5. What else are the multiplier and multiplicand called?

Factors.

REMARK.—The *product* is the *same* in whatsoever order the factors are multiplied. Thus, if 4 be represented by a horizontal row of 4 counters. and 3 by a ✦ ✦ ✦ ✦ perpendicular row of 3 counters, it is plain by ✦ ✦ ✦ ✦ inspection that the horizontal row taken 3 times ✦ ✦ ✦ ✦ is equal to the perpendicular row taken 4 times.

6. How is Multiplication denoted?

By an *oblique cross,* called the *sign of mul-'iplication* ($\times$). Thus, 3×4 shows that 3 and 4 are to e·multiplied together.

Read the following:

$2 \times 4 = 6 + 2.$ 4 times $6 = 3$ times 8.

$3 \times 6 = 2 \times 9.$ 3 times $12 = 6$ times 6.

4

1. What cost 5 oranges, at 4 cents each ?

ANALYSIS.—If 1 orange costs 4 cents, 5 oranges will cost 5 times 4, or 20 cents.   Therefore, 5 oranges will cost 20 cents.

2. What will 4 pears cost, at 3 cents each ?

3. If John writes 6 lines a day, how many lines will he write in 5 days?

4. What cost 7 writing books, at 8 cents each ?

5. At 6 dollars apiece, what cost 8 hats ?

6. What cost 9 cloaks, at 10 dollars apiece ?

7. What is the difference between 7 times 8, and 9 times 6 ?

8. If 1 orange is worth 6 apples, how many apples are 12 oranges worth ?

LESSON XIV.

When the Multiplier has but one figure, and the Product of each figure is Less than 10.

1. If 1 car costs 1304 dollars, what will 2 cars cost ?

ANALYSIS.—Write the multiplier under the *Operation.* multiplicand, and beginning at the right, proceed   1304
thus : 2 times 4 units are 8 units. Set the 8 in   2
units' place, under the figure multiplied, because   ———
it *is units.* 2 times o tens are o tens. Set the o *Ans.* 2608
in tens' place, because there are no *tens.* 2 times
3 hundreds are 6 hundreds. Set the 6 in hundreds' place, because it is *hundreds.* 2 times 1 thousand are 2 thousand. Set
the 2 in thousands' place, etc.

(2.)	(3.)	(4.)	(5.)
33114	23302	22210	30302
2	3	4	5

1. **What is the cost of 7 books, at 6 dollars each ?**

ANALYSIS.—Since 1 book costs 6 dollars, 7 books will cost 7 times 6, or 42 dollars. Therefore, 7 books cost 42 dollars.

2. What cost 6 barrels of peas, at 9 dollars a barrel ?
3. How many are 8 times 9 horses ?
4. At 7 dollars a yard, what cost 8 yards of cloth ?
5. If a pigeon flies 7 miles an hour, how far will it fly in 11 hours?
6. What cost 9 pounds of ham, at 12 cents a pound ?

LESSON XV.

When the Multiplier has but one figure, and the Product of a figure is 10 or more.

1. What will 3 pianos cost, at 536 dollars apiece ?

ANALYSIS.—Since 1 piano costs 536 dollars, 3 pianos will cost 3 times 536 dollars. Set the multiplier under the multiplicand, as before, and beginning at the right proceed thus : 3 times 6 units are 18 units, equal to 1 ten and 8 units. *Ans.* 1608 dols. Set the 8, or units' figure, under the figure by which you multiply, and add the 1 ten to the product of the next figure, as in Addition. Again, 3 times 3 are 9 tens, and 1 ten make 10 tens, equal to 1 hundred and 0 tens. Set the 0 in tens' place, and add the 1 to the product of the next figure. Finally, 3 times 5 hundreds are 15 hundreds, and 1 will make 16 hundreds. Therefore, 3 pianos will cost 1608 dollars.

Operation.
536 dols.
3
——
Ans. 1608 dols.

Multiply the following in like manner :

(2.)	(3.)	(4.)	(5.)
4568	37056	30740	64346
2	3	4	5

(6.)	(7.)	(8.)	(9.)
43025	37420	50738	69400
3	5	4	6

(10.)	(11.)	(12.)	(13.)
56036	65327	548064	437603
6	7	8	9

14. In 1 year there are 365 days: how many days are there in 14 years?   *Ans.* 5110 days.

15. If 1 house lot is worth 450 dollars, how much are 5 similar lots worth?   *Ans.* 2250 dols.

16. In 1 farm there are 320 acres: how many acres are there in 7 farms?

17. What are 6 carriages worth, at 750 dollars apiece?

18. Multiply 7203 pounds by 7.

19. Multiply 8200 dollars by 8.

20. Multiply 9345 barrels by 9.

Mental Exercises.

1. What is the price of 3 cows, at 35 dollars apiece?

ANALYSIS.—35 is the same as 3 tens and 5 units. Now 3 times 3 tens are 9 tens, or 90; and 3 times 5 units are 15 units, equal to 1 ten and 5 units, which added to 90 dollars make 105 dollars, the answer required.

2. What cost 2 melodeons, at 76 dollars apiece?

3. How many pounds are 4 times 56 pounds?

4. How many gallons are 5 times 63 gallons?

5. At 82 dollars apiece, what will be the cost of 6 wagons?

LESSON XVI.

When the Multiplier has two or more Figures.

1. What will 104 horses cost, at 245 dollars apiece ?

ANALYSIS.—Since 1 horse costs 245 dollars, 104 horses will cost 104 times 245 dollars. Write the multiplier under the multiplicand, units under units, etc., and beginning at the right as before, proceed thus: 4 times 5 units are 20 units, or 2 tens and 0 units. Set the 0, or units' figure, under the figure by which we are multiplying, and add the 2 or tens to the next product. The other figures of the

Operation.

245
104
—
980
245
——
Ans. 25480 dols.

multiplicand are multiplied by 4, and the results set down in like manner. Next, the product by 0 tens is 0; we therefore omit it.

Again, 1 hundred times 5 units are 5 hundreds. We set the 5, or units' figure, in hundreds' place, under the figure by which we are multiplying, because it is *hundreds*. The other figures of the multiplicand are multiplied, in like manner. Finally, adding these *partial products* together, the result, 25480 dollars, is the whole product required.

14. What is meant by partial products, and why so called ?

Partial Products are the *several results* which arise from multiplying the multiplicand by the respective figures of the multiplier, and are so called because they are *parts* of the entire product.

Multiply the following in like manner:

(2.)	(3.)	(4.)	(5.)
534	3215	4301	5320
23	35	42	54

LESSON XVII.

Review of Principles.

15. How write numbers for multiplication?

Write the multiplier under the multiplicand, units under units, etc.

16. When the multiplier has but *one* figure, how proceed?

Beginning at the right, multiply each figure in the multiplicand by it, and set down the result as in Addition.

17. When the multiplier has two or more figures, how?

Multiply the multiplicand by each figure of the multiplier separately, and set the first figure of each partial product under the multiplying figure. The sum of the partial products will be the answer required.

18. How is multiplication proved?

Multiply the multiplier by the multiplicand; if this result agrees with the first, the work is right.

Examples for Practice.

1. Multiply 65 by 38, and prove the operation.

Multiplicand	65	The same Multiplier	38
Multiplier	38	" " Multiplicand	65
	520		190
	195		228
Product	2470	Same result as before	2470

2.)	(3.)	(4.)	(5.)
23465	35670	48682	67086
38	57	63	75

MULTIPLICATION.

6. Allowing 320 acres to a farm, how many acres are there in 25 farms?

7. If 1 carriage is worth 850 dollars, how much are 47 carriages worth?

8. What cost 375 acres at 54 dollars an acre?

(9.)	(10.)	(11.)	(12.)
54365	45467	76734	68659
64	68	95	84

(13.)	(14.)	(15.)	(16.)
5427	6854	7496	8539
73	127	234	345

(17.)	(18.)	(19.)	(20.)
37610	74063	80945	95070
307	278	452	546

21. How many oranges will 75 trees produce, if each tree bears 2563 oranges?

22. What is the product of 563 multiplied by 153?

23. What is the product of 1275 into 206?

24. What cost 367 horses, at 305 dollars apiece?

25. What is the product of 430 multiplied by 321?

26. If a railroad car goes at the rate of 288 miles per day, how far will it go in 335 days?

(27.)	(28.)	(29.)	(30.)
3205	5023	6704	7006
246	265	405	2613

LESSON XVIII.

When the Multiplier is 10, 100, 1000, etc.

19. What is the effect of annexing a cipher to a figure?

It *removes* the figure one place to the left, and increases its value *ten times.*   P. 19, Q. 21.) Thus, 3 denotes 3 units; but 30 denotes ten times 3, or thirty.

20. What, if two ciphers are annexed?

Its value is increased a *hundred times,* or multiplied by 100.   Thus, 4 denotes 4 units, but 400 denotes a hundred times 4. or four hundred.

1. What cost 576 acres of land, at 10 dols. per acre?

2. What will 100 shawls cost, at 58 dollars each?

SOLUTION. — 58 dols. × 100 = 5800 dols. *Ans.*

21. How multiply by 10, 100, 1000, etc.?

Annex as many ciphers to the multiplicand as there are ciphers in the multiplier: the result will be the product.

NOTE. —To *annex* signifies to place *after,* or on the *right.*

3. How many days are there in 100 years, allowing 365 days to a year?

4. What is the product of 421 multiplied by 100?

5. How many are 1000 times 30348?

6. Multiply 50463 by 1000.

7. Find the product of 256 multiplied by 1000.

8. Multiply 4388 by 1000.

9. Multiply 5898 by 10000.

10. Multiply 67063 by 10000.

11. Multiply 6504 by 100000.

LESSON XIX.

When one or both factors have ciphers on the right.

1. What cost 30 city lots, at 2500 dols. apiece?

ANALYSIS.—Since 1 lot is worth 2500 dols., 2500 d.
30 lots must be worth 30 times 2500 dols., and 30
2500 dols. × 30 = 7500 dols. ———
We multiply the significant figures 25 by 3, *Ans.* 75000 d.
and to their product annex as many ciphers as
we find on the right of the multiplier and multiplicand, which
is three. The result is 75000 dols.

22. How proceed when one or both factors have ciphers on
the right?

*Multiply the significant figures together; and to the
result annex as many ciphers as are found on the right
of both factors.*

(2.)	(3.)	(4.)	(5.)
42000	3503	32000	8000
6	200	300	900

Ans. 252000   *Ans.* 9600000

6. What is the product of 9000 into 300?
7. Multiply 365 by 200.
8. Multiply 258 by 5000.
9. Multiply 3000 by 327.
10. Multiply 80000 by 94.
11. Multiply 4000 by 3000.
12. Multiply 3200 by 5400.
13. If 1 acre of land produces 40 bushels of wheat,
what will 3000 acres produce?

DIVISION.

LESSON 1.

To TEACHERS.—The design of this Lesson is to *develop* the *idea of times*, as used in *Division*, and lead the pupil to see the *similarity* of Division to Subtraction.

1. I have 2 pencils: if I hand 1 of them to you, how many shall I have left ? " One."
2. If I hand you another, how many ? " None."
3. How many times have I handed you one pencil ?
   Two times.
4. How many times can 1 pencil be taken from 2 pencils ?
5. Here are 3 pears in a fruit dish : if I take 1 of them, how many will be left ? " Two."
6. If I take 1 more. how many ? " None."
7. How many times have I taken 1 pear ?
   Three times.
8. How many times can 1 pear be taken from 3 pears ?
9. How many times 1 in 3 ? " 3 times."
10. Here are 4 apples : if I pass 1 to the first pupil, to the second, 1 to the third. and 1 to the fourth, how many times shall I have passed an apple ?
11. How many will be left ? Show it with your fingers.

LESSON II.

TO TEACHERS.—The object of this Lesson is, to illustrate the formation of the Division Table of 2.

1. Let each make two stars upon his slate as I make them upon the blackboard.   ✳ ✳

2. If you take 2 stars from 2 stars, how many will be left?   "None."

3. How many times are 2 stars contained in 2 stars?
One time.

4. Make two groups of 2 stars each; as,

✳ ✳   ✳ ✳

5. If you erase 2 of these stars, how many will be left?   "Two."

6. If you erase 2 more, how many?   "None."

7. How many times are 2 stars contained in 4 stars?
Two times.

8. Make 3 groups of 2 stars each; as,

✳ ✳   ✳ ✳   ✳ ✳

9. How many times are 2 stars contained in 6 stars?
Three times.

Let the class continue the illustration, and write out the Table, as below.

2 in 2, 1 time.	2 in 12, 6 times.
2 in 4, 2 times.	2 in 14, 7 times.
2 in 6, 3 times.	2 in 16, 8 times.
2 in 8, 4 times.	2 in 18, 9 times.
2 in 10, 5 times.	2 in 20, 10 times.

10. If Joseph can buy 1 pencil for 2 cents, how many pencils can he buy for 8 cents?

—

LESSON III.

3 in 3,	1 time.		3 in 18,	6 times.
3 in 6,	2 times.		3 in 21,	7 times.
3 in 9,	3 times.		3 in 24,	8 times.
3 in 12,	4 times.		3 in 27,	9 times.
3 in 15,	5 times.		3 in 30,	10 times.

1. How many oranges at 3 cents each can George buy for 12 cents?

ANALYSIS.—Since 3 cents will buy 1 orange, 12 cents will buy as many as 3 cents are contained times in 12 cents, which are 4 times. Therefore, he can buy 4 oranges.

2. If 3 apples are worth 1 pencil, how many pencils are 6 apples worth?

3. At 3 cents a yard, how many yards of tape can you buy for 9 cents?

4. How many threes in 15? In 12? In 9?

5. How many times can you take 3 apples from a fruit dish containing 15 apples? Show it.

6. How many times 3 make 27? 21? 24? 18?

7. What is finding how many times one number is contained in another called?

Division.

8. What is the result obtained by division called?

The Quotient.

Copy and divide the following:

(9.)	(10.)	(11.)	(12.)	(13.)	(14.)
3)21	3)24	2`14	2`16	3`27	3`30

LESSON IV.

4 in 4,	1 time.	4 in 24,	6 times.
4 in 8,	2 times.	4 in 28,	7 times.
4 in 12,	3 times.	4 in 32,	8 times.
4 in 16,	4 times.	4 in 36,	9 times.
4 in 20,	5 times.	4 in 40,	10 times.

1. A teacher having 12 apples, divided them equally among 4 pupils: how many did each receive?

ANALYSIS.—Since 12 apples were divided equally among 4 pupils, each pupil received as many apples as 4 is contained times in 12; and 4 is in 12, 3 times. Therefore, each pupil received 3 apples.

2. If you give 8 marbles to 4 boys, how many will each receive?

3. Henry paid 16 cents for 4 pears: how much was that apiece?

4. If you divide 20 pounds of flour equally among 4 persons, how many pounds will each receive?

5. How many fours in 16? In 20? In 24?

6. Sold 4 vests for 28 dollars: what was that apiece?

7. A teacher bought 4 slates for 40 cents: what was the cost of each?

8. If you pay 36 cents for 4 quarts of chestnuts, what will that be a quart?

9. How many fours in 36? In 24? In 32?

Copy and divide the following:

(10.)	(11.)	(12.)	(13.)	(14.)	(15.)
4 20	4 32	4`28	4)36	4`24	4 40

LESSON V.

5 in 5,	1 time.	5 in 30,	6 times.
5 in 10,	2 times.	5 in 35,	7 times.
5 in 15,	3 times.	5 in 40,	8 times.
5 in 20,	4 times.	5 in 45,	9 times.
5 in 25,	5 times.	5 in 50,	10 times.

1. If you divide 10 marbles into 5 equal parts, how many will there be in each part?

2. If you pay 15 cents for 5 yards of tape, how much will that be a yard?

3. Bought 5 quarts of milk for 20 cents: what was that a quart?

4. If 30 children are seated equally on 5 benches, how many will there be on a bench?

5. If 5 guns cost 40 dollars, what will 1 gun cost?

6. A man having 35 acres of land, fenced it into 5 equal pastures: how many acres to a pasture?

7. If you walk 45 miles in 5 days, how many miles will you walk per day?

8. A teacher distributed 50 dollars in equal prizes, among 5 of his best pupils: what did each receive?

9. A young man paid 40 dollars for boarding 8 weeks: what was that per week?

10. George received 45 cents for picking 5 quarts of strawberries: what was that a quart?

Copy and divide the following:

(11.)	(12.)	(13.)	(14.)	(15.)
5)25	5)30	5)45	5)35	5)50

LESSON VI.

6 in 6,	1 time.	6 in 36,	6 times.
6 in 12,	2 times.	6 in 42,	7 times.
6 in 18,	3 times.	6 in 48,	8 times.
6 in 24,	4 times.	6 in 54,	9 times.
6 in 30,	5 times.	6 in 60,	10 times.

1. How many yards of ribbon at 6 cents a yard can you buy for 18 cents?

2. If you put 6 oranges in a basket, how many baskets will be required to hold 36 oranges?

3. If I pay 12 dollars for 6 books, what will the books cost me apiece?

4. In 6 pages the printer made 24 mistakes: how many mistakes was that to a page?

5. How many quarts of milk, at 6 cents a quart, can be had for 36 cents?

6. At 6 dollars apiece, how many muffs can be purchased with 42 dollars?

7. How many times can you take 6 marbles from 48 marbles?

8. If Samuel gains 6 credits each day, how many days will it take him to gain 54 credits?

9. If a man lays up 6 dollars a week, how long will it take him to lay up 60 dollars?

Copy and divide the following:

(10.)	(11.)	(12.)	(13.)	(14.)	(15.)
6)30	6)24	6)42	6)48	6)60	6)54

LESSON VII.

7 in 7,	1 time.	7 in 42,	6 times.
7 in 14,	2 times.	7 in 49,	7 times.
7 in 21,	3 times.	7 in 56,	8 times.
7 in 28,	4 times.	7 in 63,	9 times.
7 in 35,	5 times.	7 in 70,	10 times.

1. If 7 yards of twist are worth 21 cents, what is 1 yard worth ?

2. If Louise reads 28 pages in 7 days, how many pages will she read in 1 day ?

3. In 35 lemons, how many times 7 lemons?

4. In 7 days there is 1 week : how many weeks in 42 days ?

5. If Sanford shoots 35 squirrels in 7 days, how many will he shoot in 1 day ?

6. How many sevens in 63 ? How many in 56 ? In 28 ? In 35 ? In 70 ?

7. If 7 pounds of sugar cost 56 cents, how much will 1 pound cost ?

8. If 7 loaves of bread will last a family 1 week, how many weeks will 49 loaves last ?

9. In a peach orchard there are 63 trees with 7 trees in a row : how many rows are there ?

10. How many times 7 pencils make 56 pencils?

11. If 7 dozen eggs cost 70 cents, what cost 1 dozen ?

Copy and divide the following :

(12.)	(13.)	(14.)	(15.)	(16.)	(17.)
7)21	7)35	7)49	7)56	7)63	7)70

LESSON VIII.

8 in 8, 1 time.	8 in 48, 6 times.
8 in 16, 2 times.	8 in 56, 7 times.
8 in 24, 3 times.	8 in 64, 8 times.
8 in 32, 4 times.	8 in 72, 9 times.
8 in 40, 5 times.	8 in 80, 10 times.

1. At 8 cents apiece, how many writing-books can be had for 24 cents?

2. How many times can you take 8 chairs from a row of 32 chairs?

3. In 8 quarts there is 1 peck: how many pecks are there in 40 quarts?

4. How many eights are in 56? In 48? In 64?

5. How many classes of 8 can be formed of 72 pupils?

6. Paid 48 cents for 6 slates: what is that apiece?

7. A laundress paid 64 cents for 8 pounds of starch: what was that a pound?

8. How many times can you draw 8 quarts of vinegar from a cask containing 80 quarts?

9. How many 8-pound rolls can be made from a keg of butter containing 56 pounds?

10. William picked 8 quarts of blackberries, and sold them for 48 cents: what was that a quart?

11. How many times 8 in 32? In 24? In 72?

Copy and divide the following:

(12.)	(13.)	(14.)	(15.)	(16.)	(17.)
8)32	8)48	8)64	8)56	8)72	8)80

LESSON IX.

9 in 9,	1 time.	9 in 54,	6 times.
9 in 18,	2 times.	9 in 63,	7 times.
9 in 27,	3 times.	9 in 72,	8 times.
9 in 36,	4 times.	9 in 81,	9 times.
9 in 45,	5 times.	9 in 90,	10 times.

1. A teacher having 27 pupils, arranged them in 3 equal classes: how many were in each class?

2. At 9 dollars apiece, how many vests can be had for 36 dollars?

3. A farmer planted 54 peach trees in 9 equal rows: how many did he put in a row?

4. Joseph caught 45 fish in 9 hours: how many was that per hour?

5. A farmer gathered 63 bushels of apples from 9 trees: how many bushels did this average a tree?

6. How many times can 9 be taken from 36? From 45? From 54? From 63?

7. If you pay 9 cents a quart for blackberries, how many quarts can you buy for 72 cents?

8. If a hunter catches 9 pigeons at a time, how many times must he spring his net to catch 81 pigeons?

9. At 9 cents a pack, how many packs of fire-crackers can be bought for 90 cents?

10. How many nines in 45? In 36? In 54? In

(11.)	(12.)	(13.)	(14.)	(15.)	(16.)
9)36	9)54	9)45	9)63	9)81	9)72

LESSON X.

10 in 10,	1 time.	10 in 60,	6 times.
10 in 20,	2 times.	10 in 70,	7 times.
10 in 30,	3 times.	10 in 80,	8 times.
10 in 40,	4 times.	10 in 90,	9 times.
10 in 50,	5 times.	10 in 100,	10 times.

1. Sarah paid 40 cents for ribbon, which was 10 cents a yard : how many yards did she buy ?

2. In 10 cents there is 1 dime : how many dimes are there in 60 cents ?

3. If you pay 10 cents a mile for a horseback ride, how many miles can you ride for 50 cents?

4. If 10 chestnuts are worth 1 orange, how many oranges are 30 chestnuts worth ?

5. At 10 dollars each, how many accordeons can be purchased for 60 dollars?

6. How much hay, at 10 dollars a ton, can be bought with 80 dollars?

7. How long will it take a horse to go 60 miles, if he goes 6 miles an hour ?

8. Ten lads paid 90 cents for the use of a sail-boat : how much was that apiece?

9. Ten dollars make 1 eagle : how many eagles in 100 dols.

10. How many tens in 40? In 50? In 90?

Copy and divide the following:

(11.)	(12.)	(13.)	(14.)	(15.)	(16.)
10)60	10)50	10)70	10)90	10)80	10)100

LESSON XI.

Explanation of Terms.

1. What is Division?

Division is finding how many times one number is contained in another.

2. What is the number to be divided, called?

The *Dividend.*

3. The number to divide by?

The *Divisor.*

4. What is the number obtained by division, called?

The *Quotient.*

5. What is the number *left*, called?

The *Remainder.*

When it is said that 2 is contained in 9, 4 times and 1 over, which is the dividend? The divisor? The quotient? The remainder?

REMARK.—A *proper* remainder is always *less* than the divisor.

6. How is Division denoted?

By a *short horizontal line* between two dots (÷), called the *Sign of division.*

7. When placed between two numbers, what does it show?

It shows that the number *before* it is to be divided by the one *after it.* Thus, $15 \div 3$, shows that 15 is to be divided by 3, and is read, " 15 divided by 3."

8. How else is division denoted?

By *writing the divisor under the dividend* with a short line between them; as $\frac{15}{3}$.

Copy and read the following: $8 \div 2 = 4$; $18 \div 3 = 4 + 2$; $28 \div 4 = 4 + 3$; $3 + 2 = 20 \div 4$; $\frac{21}{3} = 7 + 2$; $\frac{32}{4} = 6 + 2$.

LESSON XII.

To Teachers.—The design of this Lesson is to illustrate the *two objects*, or *classes of examples*, to which Division is applied.

1. Charles has 6 cents to buy apples which are 2 cents each : how many can he buy?

Analysis.—Since 2 cents will buy 1 apple, 6 cents will buy as many apples as 2 cents are contained times in 6 cents. The object here is to find how many times one number is contained in another.

Let the 6 cents be represented by 6 counters; as.

✦ ✦ ✦ ✦ ✦ ✦

Separating them into groups of 2 counters each, we have 3 groups. Therefore, he can buy 3 apples.

2. James has 6 oranges to be divided equally between his 2 brothers: how many will each receive?

Analysis.—Since 2 brothers are to receive 6 oranges, each will receive 1 orange as often as 2 is contained in 6. The object here is to divide 6 oranges into 2 equal parts, and find how many there are in each part.

Let the 6 oranges be represented by 6 counters; as,

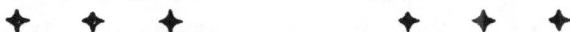

✦ ✦ ✦ ✦ ✦ ✦

Separating them into two groups, putting 1 counter in each, then another, and so on, till the counters are exhausted, we have 3 counters in each group. Therefore, each brother will receive three oranges.

9. To what two classes of examples is Division applied?

First. Those in which it is required to *find how many times one number is contained in another.* (See examples, pp. 83, 84.)

Second. Those in which it is required to *divide a number into equal parts.* (See examples, pp. 85, 86.)

LESSON XIII.

NOTE.—In the *First* class of examples referred to, the learner will observe that the divisor and dividend are the *same denomination*, and the quotient is *times*, or an *abstract* number.

In the *Second*, the divisor and dividend are *different denominations*, and the *quotient* is the same denomination as the *dividend*. The process of reasoning in their solution is somewhat different: but the practical operation is the same, viz.: *to find how many times* one number is contained in another, which accords with the definition of Division.

1. A farmer divided 30 pounds of maple sugar equally among 6 lads: what part, and how much did 1 lad receive?

ANALYSIS.—1 is 1 sixth of 6; therefore, 1 lad received 1 sixth of 30 pounds; and 1 sixth of 30 pounds is 5 pounds.

2. At 5 cents each, how many whistles can be bought for 20 cents?

ANALYSIS.—Since 5 cents will buy 1 whistle, 20 cents will buy as many as 5 cents are contained times in 20 cents, which is 4 times. Therefore, 20 cents will buy 4 whistles.

3. To which class does Ex. 1 belong? Why? Ex. 2? Why?

4. If 6 men can earn 18 dollars per day, how much can 1 man earn in the same time?

5. At 8 cents a quart, how much milk can be had for 56 cents?

6. If a teacher having 60 pupils puts 12 in a class, how many classes will he have?

7. How many drums, at 9 dollars each, can be bought for 108 dollars?

LESSON XIV.

Slate Exercises in Short Division.

1. How many apples, at 2 dollars a barrel, can you obtain for 4602 dollars?

ANALYSIS.—Write the divisor on the left of *Operation* the dividend, with a curve line between them, and 2)4602 proceed thus: 2 is contained in 4, 2 times; write ——— the 2 under the figure divided, for it is the *same Ans.* 2301 b. *order* as that figure. Next, 2 is contained in 6, 3 times; write the 3 under the figure divided, for the same reason. 2 is contained in 0, no times; write a cipher in the quotient. 2 is in 2, 1 time; set the 1 under the figure divided.

(2.)   (3.)   (4.)   (5.)
2)6420   3 6309   4 8048   5)5005

(6.)   (7.)   (8.)   (9.)
6)5606   7)7070   8)8008   9)9090

Mental Exercises.

1. How many balls, at 8 cents, can Henry buy for 50 cents, and how many cents will he have left?
2. If you devide 54 chestnuts equally among 6 of your companions, what part and how many will each receive?
3. How many times is 7 contained in 38, and how many over?

LESSON XV.

When the Divisor is not contained exactly in each Figure of the Dividend.

1. How many yards of cloth, at 4 dollars a yard, can a tailor buy for 126027 dollars?

ANALYSIS.—Since he can buy 1 yard for 4   *Operation.*
dollars, for 126027 dollars he can buy as many   4)126027
yards as 4 is contained times in 126027.   ————————
Write the numbers and divide as before.   *Ans.* 31506¾ y
But the divisor is not contained in the first
figure of the dividend, therefore we find how many times it is
contained in the first two. Thus, 4 is in 12, 3 times. Set the 3
under the right hand figure divided. Again, 4 is in 6, 1 time
and 2 remainder. Set the 1 under the figure divided, and prefixing the remainder mentally to the next figure, makes 20.
Now, 4 is in 20, 5 times. Set the 5 under the figure divided.
Next, 4 is not contained in 2, we therefore put a cipher in
the quotient, and prefix the 2 to the next figure, as if a remainder. making 27. Now 4 is in 27, 6 times and 3 remainder.
Writing this last remainder over the divisor, we annex it to the
quotient.

Copy and divide the following in a similar manner:

(2.)	(3.)	(4.)	(5.)
3)2045	4)56368	6)15006	6)84506
681⅔		3001¼	

(6.)	(7.)	(8.)	(9.)
3)43000	4)78400	5)60903	6)84500
Ans. 14333⅓		12180⅗	

LESSON XVI.

Review of Principles.

10. When the results are carried in the mind, and the *quotient* only is set down, what is the operation called ?

Short Division.

11. How divide by *Short Division ?*

Place the divisor on the left of the dividend, and beginning at the left, divide each figure by it, setting the result under the figure divided.

12. If the divisor is not contained in a figure of the dividend, how proceed ?

Put a cipher in the quotient, and find how many times the divisor is contained in the next two figures.

13. If a remainder arises from any figure before the last, how proceed ?

Prefix it mentally to the next figure, and divide as before.

14. If from the last, how ?

Place it over the divisor, and annex it to the quotient.

15. How is Division proved ?

Multiply the divisor and quotient together, and to the product add the remainder. If the result is equal to the dividend, the work is right.

Examples for Practice.

1. How many books, at 5 dollars each, can be obtained for 18038 dollars ?   *Ans.* 3607 and 3 over.

5

(2.)	(3.)	(4.)	(5.)
2)23048	3)27506	4)60745	5)41378

(6.)	(7.)	(8.)	(9.)
6)37434	7)42359	8,41605	9)56608

10. How long will it take Julia to braid 560 straw hats, if she braids 4 each day?

11. If a man walks 3 miles an hour, how long will it take him to go 1000 miles?

12. Into how many fields can a farm of 360 acres be divided, each field containing 6 acres?

13. At 8 dollars a barrel, how many barrels of flour can be had for 640 dollars?

14. In 1 week there are 7 days: how many weeks are in 365 days, and how many days over?

15. Divide 54672 by 3.   16. Divide 45060 by 5.
17. Divide 60456 by 4.   18. Divide 72036 by 6.
19. Divide 46075 by 7.   20. Divide 66408 by 8.
21. Divide 81378 by 9.   22. Divide 90457 by 10.

23. A carpenter built 5 houses at a cost of 16550 dollars: what was the cost of each?

24. A miner sent 40260 pounds of coal to market in 6 cars: how many pounds was that to a car?

25. A man bought a farm for 5200 dollars, and paid for it in 4 annual payments: how much did he pay each year?

26. How many miles must a bird fly per day, to make 2600 miles in 8 days?

27. If a man spends 10 dollars a day, how long will it take him to spend 3650 dollars?

LESSON XVII.

Slate Exercises in Long Division.

1. Divide 14120 by 3, using Long Division.

First. Find how many times the divisor is contained in the first, or first two figures on the left of the dividend, and set the quotient 4 on the right, with a curve line between them.

Second. Multiply the divisor by this quotient figure, and set the product 12 under the figures divided.

Third. Subtract this product from the figures divided. *Fourth.* Annex to the remainder 2, the next figure of the dividend, making 21 for a new partial dividend.

Div. Div'd. Quotient.

```
3)14120(4706⅔
  12
  —
  21
  21
  —
  020
   18
  —
  2 Rem.
```

Dividing, etc., as before, the third partial dividend is 2. But the divisor 3 is not contained in 2. We therefore place a cipher in the quotient, and bringing down the next figure, divide as before. Finally, placing the remainder arising from the last figure over the divisor and annexing it to the quotient —the result is 4706⅔.

NOTE.—To prevent mistakes, it is customary to place a *mark* under the several figures of the dividend, when brought down.

Copy and divide the following by Long Division.

(2.)	(3.)	(4.)	(5.)
4)3328	3)3564	5)4570	6)5256

(6.)	(7.)	(8.)	(9.)
6)45735	8)56450	7)65384	9)87845

LESSON XVIII.

Review of Principles.

16. When the *results* of the several steps and the *quotient* are all set down, what is the operation called?

Long Division.

17. What is the first step in Long Division?

Find how many times the divisor is contained in the fewest figures on the left of the dividend that will contain it.

18. The second?

Multiply the divisor by the quotient figure, and set the product under the figures divided.

19. The third?

Subtract the product from the figures divided.

20. The fourth?

Annex to the remainder the next figure of the dividend, for a new partial dividend; then divide as before.

21. What is to be done with the final remainder?

Set it over the divisor, and annex it to the quotient.

NOTES.—1. If the *product* of the divisor into the figure placed in the quotient is *greater* than the partial dividend, the quotient figure is *too large*, and therefore must be *diminished*.

2. If the *remainder* is *equal* to or *greater* than the *divisor*, the quotient figure is *too small*, and must be *increased*.

10. How many times is 15 contained in 10523?

ANALYSIS.—Since the divisor is not contained in the first two figures of the dividend, we find how many times it is contained in the first three, the fewest that will contain it; then multiply, etc.

Operation.

$15)10523(701 \tfrac{8}{15}$
$\underline{105}$
$\quad 02$, etc.

1. Divide 1506 by 23.   *Ans.* 65$\frac{11}{23}$.
2. Divide 2536 by 8.
3. Divide 3745 by 12.
4. Divide 4678 by 15.
5. Divide 5168 by 9.
6. Divide 6238 by 25.

7. Divide 34568 by 24.
8. Divide 44605 by 21.
9. Divide 66431 by 32.
10. Divide 75054 by 37.
11. Divide 96387 by 45.

12. A drover laid out 5600 dollars in cattle, at 25 dollars a head: how many did he buy?

13. If a man can earn 36 dollars a month, how long will it take him to earn 432 dollars?

14. A man invested 9765 dollars in land, at 20 dollars an acre: how many acres did he buy?

15. Allowing 63 gallons to a hogshead, how many hogsheads can be filled from a cistern holding 7000 gallons?

16. How long will 15000 dollars support a person whose expenses are 75 dollars a week?

17. A captain distributed 3150 pounds of flour among a company of soldiers, giving each 45 pounds: how many were in his company?

18. Allowing 52 weeks to a year, how many years in 5252 weeks?

19. At 95 dollars apiece, how many horses can be purchased for 9800 dollars?

20. If a person travels 58 miles a day, how long will it take him to travel 3480 miles?

21. Divide 80045 by 61.   23. Divide 85784 by 63.
22. Divide 75007 by 56.   24. Divide 90705 by 75.

LESSON XIX.

To find the Quotient Figure when the Divisor is large.

1. Divide 12328 by 382.

ANALYSIS.—If we take 3 for a trial divisor, it is contained in 12, 4 times. But multiplying 8 by 4, we have 3 to add to the product of the next figure, and 3 added to 4 times 3, make 15, which is larger than 12, the figures divided. Hence 4 is too large for the quotient figure. We therefore place 3 in the quotient, and proceed as before.

Operation.

```
382)12328(32
    1146
    ————
     868
     764
    ————
Rem. 104
```

22. How find the quotient figure, when the divisor is large ?

Take the first figure of the divisor for a trial divisor, and find how many times it is contained in the first or first two figures of the dividend, making due allowance for carrying the tens of the product of the second figure of the divisor into the quotient figure.

2. Divide 53643 by 213.   5. Divide 678543 by 503.
3. Divide 64078 by 345.   6. Divide 724608 by 625.
4. Divide 43840 by 456.   7. Divide 830245 by 2345.

LESSON XX.

When the Divisor is 10, 100, 1000, etc.

1. How many horses, at 100 dollars apiece, can be bought for 1545 dollars ?

ANALYSIS. — *Removing* a cipher from the *right* of a number, *divides* it by 10 ; for, each figure in the number is removed one place to the right. (P. 19, Q 12.)

Operation.

1'00)15'45

Quo't, 15. 45 *Rem.*

In like manner, cutting off *two* figures from the right of a number, divides it by 100; cutting off *three*, by 1000, etc.

As the divisor is 100, it is only necessary to cut off two figures on the right of the dividend; those left, are the quotient, and those cut off, the remainder.

23. How proceed when the divisor is 10, 100, 1000, etc.?

From the right of the dividend cut off as many figures as there are ciphers in the divisor. The figures left will be the quotient; those cut off, the remainder.

2. Divide 564 by 100.   5. 39467 by 10000.
3. Divide 6531 by 1000.   6. 72364 by 100000.
4. Divide 8000 by 1000.   7. 200000 by 100000.

8. Divide 2354 by 20.

ANALYSIS.—We cut off the cipher on the right of the divisor, and the figure 4 on the right of the dividend; then divide by 2, the other figure of the divisor. The result, 117, is the quotient, and 4, the figure cut off, being annexed to the remainder 1, is the true remainder.

Operation.
2|0)235|4
———
Quo't, 117,14 *Rem.*

24. How proceed, when the divisor is composed of significant figures, with ciphers on the right?

I. *Cut off the ciphers on the right of the divisor, and as many figures on the right of the dividend.*

II. *Divide the remaining part of the dividend by the remaining part of the divisor for the quotient.*

III. *Annex the figures cut off to the remainder, and the result will be the true remainder.*

9. Divide 6533 by 20.   11. Divide 43681 by 210.
10. Divide 42345 by 30.   12. Divide 48642 by 2300.

LESSON XXI.

Oral Drill.

To Teachers.—Better results will be obtained by Oral Drills, *short*, *spirited*, and *frequent*, than from scores of examples recited in a *sluggish*, *slipshod* manner.

1. To 3 add 4; subtract 2; multiply by 4; divide by 2; subtract 4; multiply by 5; divide by 10: result?

EXPLANATION.—The teacher names the number and the operation to be performed slowly at first; the class perform the operation mentally. Thus, the teacher says, "to 3 add 4," the class think 7; "subtract 2;" the class think 5; "multiply by 4;" the class think 20, and so on.

2. From 8 take 5; add 3; multiply by 4; divide by 6; multiply by 5; take 4; divide by 2: result?

3. Multiply 4 by 9; divide by 6; add 7; subtract 3; divide by 5; multiply by 8; add 4: result?

4. Divide 15 by 3; multiply by 6; take 3; divide by 9; multiply by 8; add 10; take 5: result?

5. From 15 take 7; multiply by 6; divide by 8; add 12; take 3; add 9; divide by 6: result?

6. To 14 add 6; divide by 5; multiply by 8; take 5: divide by 3; add 3; multiply by 5: result?

7. Multiply 7 by 6; add 8; divide by 10; multiply by 7; take 5; divide by 3; multiply by 8: result?

8. Divide 45 by 9; multiply by 8; take 10; divide by 6; multiply by 7; add 10; add 8: result?

9. From 25 take 7; divide by 6; multiply by 9; add 9; divide by 12: multiply by 10: divide by 6; multiply by 20; divide by 10; add 30; divide by 8: result?

FRACTIONS.

LESSON I.

1. If a unit is divided into two equal parts, what is each part called?

One half.

2. If divided into three equal parts, what are the parts called?

Thirds.

3. If divided into four equal parts, what are the parts called?

Fourths.

1. Draw a line the length of your book, and divide it into halves. Into thirds. Into fourths.

2. If divided into seven equal parts, what are the parts called? If divided into ten, what? Into twelve?

3. What is meant by a half? (See Q. 1.)

4. What is meant by a third? Two thirds? A fourth? Three fourths? By fifths? Eighths?

5. How many halves make a unit or one? How many thirds? Fourths? Sixths? Tenths?

6. What is a half of 2 cents?

ANALYSIS.—If you divide 2 cents into 2 equal parts, one of these parts is 1 cent.

7. What is a half of 4? Of 6? Of 8? 10? 12?

LESSON II.

1. What part of 2 is 1 ?   "One half."
2. What part of 3 is 1 ?   "One third."
3. What part of 3 is 2 ?

ANALYSIS.—Two is two times 1; therefore, two is 2 times 1 third, or *two thirds* of 3.

4. What part of 4 is 1 ?   Is 2 ?   Is 3 ?
5. What part of 5 is 1 ?   Of 8 is 3 ?   Of 10 is 7 ?
6. If I divide 6 pencils equally among 3 pupils, what part, and how many, will each receive?

ANALYSIS.—One is 1 third of 3. Therefore, each will receive 1 third part. And 1 third part of 6 pencils is 2 pencils.

7. If I divide 10 apples equally among 5 girls, what part, and how many, will each receive?
8. How many halves in two apples?

ANALYSIS.—In 1 apple there are two halves, and in 2 apples there must be 2 times 2 halves, which are 4 halves.

9. How many halves in 4 ?   In 5 ?   In 6?   In 10?
10. How many halves in 3 and 1 half ?

ANALYSIS.—Since in one there are 2 halves, in 3 there are 3 times 2 or 6 halves, and 1 half will make 7 halves.

11. How many thirds in 4 and 2 thirds?
12. How many fifths in 4 and 3 fifths?
13. How many units in 4 halves?

ANALYSIS.—Since in 2 halves there is 1 unit, in 4 halves there are as many units as 2 halves are contained times in 4 halves, which is 2 times. *Ans.* 2 units.

14. How many units in 6 halves?   In 9 thirds?   In 8 fourths?

LESSON III.

Explanation of Terms.

4. When a number or thing is divided into equal parts, what are the parts called?

Fractions.

5. From what do these parts take their name?

From the *number of equal parts* into which the unit is divided.

6. How are fractions commonly expressed?

By *figures* written above and below a line, called the *numerator* and *denominator;* as $\frac{2}{3}$, $\frac{3}{4}$, $\frac{7}{12}$.

7. Where is the Denominator placed, and what show?

The **Denominator** is written *below* the line, and shows into *how many equal parts* the unit is divided.

8. Where the Numerator, and what does it show?

The **Numerator** is written *above* the line, and shows *how many parts* are expressed by the fraction.

1. Express the following fractions by figures: One-half, Two-thirds, One-fourth, Three-fourths, Four-fifths, Five-eighths, Nine-tenths, Five-thirds, Seven-fourths, Four-fourths, Twelve-fifteenths, Nineteen-twentieths.

Copy and read the following fractions:

(2.)	(3.)	(4.)	(5.)	(6.)	(7.)
$\frac{2}{3}$	$\frac{2}{3}$	$\frac{5}{7}$	$\frac{4}{3}$	$\frac{9}{10}$	$\frac{4}{4}$
$\frac{3}{4}$	$\frac{4}{5}$	$\frac{3}{7}$	$\frac{5}{4}$	$\frac{7}{7}$	$\frac{9}{10}$
$\frac{2}{5}$	$\frac{5}{8}$	$\frac{8}{10}$	$\frac{7}{15}$	$\frac{13}{13}$	$\frac{23}{15}$

LESSON IV.

1. If George has 3 half dollars, and his father gives him 2 halves more, how many dollars will he have?

ANALYSIS.—3 halves and 2 halves are 5 halves; and 5 half dollars are equal to 2 and 1 half dollars.

2. How many are 3 fourths and 5 fourths? $\frac{3}{4}$ and $\frac{5}{4}$?

3. How many are 4 fifths and 8 fifths? $\frac{4}{5}$ and $\frac{8}{5}$?

4. The price of a top is 2 and 1 half cents, and that of a whistle 3 and 1 half cents: what is the price of both?

ANALYSIS.—2 cents and 3 cents are 5 cents, and 1 half and 1 half are 2 halves, or 1 cent, which added to 5 makes 6 cents.

5. How many dollars are 3 dollars and 3 fourths added to 5 dollars and 2 fourths?

6. If you have a pie and give away 3 fourths of it, how much will you have left?

ANALYSIS.—In 1 pie there are 4 fourths; and 3 fourths from 4 fourths leave 1 fourth.

7. 3 fifths from 7 fifths leave how many? $\frac{3}{5}$ from $\frac{7}{5}$?

8. 5 eighths from 9 eighths leave how many?

9. Henry's kite line was 5 and 3 fourths yards long, and he lost 2 and 1 fourth yards: how long was the part left?

ANALYSIS. — 2 yards from 5 yards leave 3 yards, and 1 fourth from 3 fourths leaves 2 fourths. Therefore, the part left was 3 and 2 fourths yards long.

10. If you have 7 eighths of a dollar, and spend 3 eighths, how much will you have left?

LESSON V.

1. What will 4 apples cost, at 1 half cent each?

ANALYSIS.—Since 1 apple costs 1 half cent, 4 apples will cost 4 times 1, or 4 half cents: and 4 half cents equal 2 cents.

2. What cost 3 plums, at 1 third cent apiece?

3. What cost 8 marbles, at 1 fourth cent apiece?

4. What cost 5 pencils, at 2 and 1 half cents apiece?

ANALYSIS.—Since 1 pencil costs 2 and 1 half cents, 5 pencils will cost 5 times as much. Now 5 times 2 cents are 10 cents, and 5 times 1 half cent are 5 halves, equal to 2 and 1 half cents, which added to 10 cents make 12 and 1 half cents.

5. James bought 6 oranges, at 4 and 1 half cents each: what did they come to?

6. How many cents are 4 times 6 and 1 fourth cents?

7. How many are 8 times 5 and 3 fourths?

8. How many are 7 times 4 and 3 fifths? 6 times 2⅓?

9. If a pear costs 4 cents, what will 1 half a pear cost?

ANALYSIS.—If 1 pear costs 4 cents, 1 half a pear will cost 1 half of 4 cents, which is 2 cents.

10. If you pay 12 cents for a pie, what must you pay for 1 third of a pie? What is $\frac{1}{3}$ of 12?

11. At 10 dollars a barrel, what cost 1 fifth of a barrel of flour? What is $\frac{1}{5}$ of 15?

12. Charles gave 3 marbles to his brother, which were 1 half of all he had: how many had he?

ANALYSIS.—Since 3 marbles are 1 half the number, 2 halves (or the whole number) must be 2 times 3 marbles, which are 6 marbles.

LESSON VI.

1. If 2 pounds of butter cost $\frac{4}{5}$ of a dollar, what will 1 pound cost?

ANALYSIS.—1 is 1 half of 2, therefore, 1 pound will cost 1 half as much as 2 pounds, and 1 half of $\frac{4}{5}$ is $\frac{2}{5}$ of a dollar.

2. If 3 yards of muslin cost $\frac{6}{8}$ of a dollar, what will 1 yard cost? What is $\frac{1}{3}$ of $\frac{6}{8}$?

3. If you can buy a pear for 3 cents, what part of a pear can you buy for 1 cent?

4. At 6 dollars a yard, what part of a yard of cloth can you buy for 1 dollar? For 3 dollars?

5. At 4 cents a yard, how much elastic cord can I buy for 5 cents?

ANALYSIS.—Since 4 cents will buy 1 yard, 1 cent will buy 1 fourth of a yard; and 5 cents will buy 5 times 1 fourth, which are 5 fourths, or 1 yard and 1 fourth.

6. At 8 cents a pound, how much maple sugar can be had for 17 cents?

7. At 6 dollars a barrel, how much flour can be bought for 27 dollars?

8. If 1 yard of braid costs 6 cents, what will $\frac{2}{3}$ of a yard cost?

ANALYSIS.—Since 3 thirds (1 yard) cost 6 cents, 1 third will cost 1 third of 6 cents, which is 2 cents. Again, if 1 third of a yard costs 2 cents, 2 thirds will cost 2 times 2, or 4 cents.

9. If 1 pound of almonds costs 25 cents, how much will $\frac{3}{5}$ of a pound cost?

10. What will $\frac{3}{4}$ of a barrel of cranberries come to, at 12 dollars a barrel?

Gold Coins.

Silver Coins.

Minor Coins.

Nickel.   Bronze.

UNITED STATES MONEY.

1. What are the denominations of U. S. money?
Eagles, dollars, dimes, cents and mills.
2. Recite the TABLE.

10	mills (*m.*)	are	1 cent,	*ct.*
10	cents	"	1 dime,	*d.*
10	dimes	"	1 dollar,	*dol.* or **$.**
10	dollars	"	1 eagle,	*E.*

100 cents = 1 dollar;   50 cents = $\frac{1}{2}$ dollar.

25 cents = $\frac{1}{4}$ dollar;   12$\frac{1}{2}$ cents = $\frac{1}{8}$ dollar.

3. How are the coins of the U. S. divided ?

Into *gold* coins, *silver* coins, and the *minor* coins.

4. Name the coins of each.

1. The *gold coins* are the *double-eagle*, the *eagle*, *half-eagle*, *quarter-eagle*, the *dollar* and *three-dollar pieces*.

2. The *silver coins* are the " trade " *dollar, half-dollar, quarter-dollar, twenty-cent piece*, and *dime*.

3. The *minor coins* are the nickel *five-cent* and *three-cent pieces*, and the *bronze cent*.

1. How many cents in 3 dimes?

ANALYSIS.—Since there are 10 cents in 1 dime, in 3 dimes there are 3 times 10 cents, which are 30 cents.

2. How many cents in 4 dimes?   In 6 dimes?

3. How many cents in 8 dols. ?   In 10 dols. ?

4. How many dimes in 40 cents?

ANALYSIS.—Since there is 1 dime in 10 cents, in 40 cents there are as many dimes as 10 cents are contained times in 40 cents, which is 4 times.   *Ans.* 4 dimes.

5. How many dimes in 20 cents?   In 50 cents?

6. How many dollars in 20 dimes?   In 40 dimes ?

7. How many dollars in 250 cents?   In 430 cents?

5. How is U. S. money expressed?

Dollars are written on the left of the decimal point, with the dollar mark before them ; cents, in the first two places on the right ; mills in the next, and parts of a mill in the succeeding places toward the right.

NOTES.—1. Eagles are expressed by *tens of dollars*, and dimes by *tens of cents*.   Thus, 5 eagles are written as $50, and 6 dimes as 60 cents.

2. As *cents* occupy *two* places, if the number to be expressed is *less* than 10, a *cipher* must be prefixed to it.

8. Write twenty dols., twenty cts., and 3 mills.

Ans. $20.203.

9. Write fifty-six dols., thirty-seven cts., six mills.

6. How is U. S. money read?

Read the figures on the left of the decimal point as dollars; those in the first two places on the right as cents; the next as mills; the others as parts of a mill.

10. Copy and read $78.35 ; $106.37 ; $308.735 ; $430.064; $57240; $60200.

ADDITION OF U.S. MONEY.

11. What is the sum of $20.465; $6.28; and $12?

ANALYSIS.—We write dollars under dollars, cents under cents, etc., and add in the usual way, placing the *decimal point* in the amount under those in the numbers added. (P. 112, Q. 5.)

$$\begin{array}{r} \$20.465 \\ 6.28 \\ 12.00 \\ \hline \$38.745 \end{array}$$

7. How add U. S. money?

Write dollars under dollars, cents under cents, etc., and add as in simple numbers, placing the decimal point in the amount under those in the numbers added.

NOTE.—If any of the given numbers have *no cents*, their place should be supplied by *ciphers*.

(12.)	(13.)	(14.)	(15.)
342.375	460.508	653.36	846.258
45.26	503.40	40.457	6.07
840.385	675.235	892,295	75.345

16. What is the sum of $375.408; $8.254; $235; and $100.40?

17. A man paid $5 625 for a hat, $18.50 for a coat, and $7.375 for a vest: what did he pay for all?

18. A lady bought a pair of gloves for $1.75, a fan for $0.87 cents, and a parasol for $6.25: what did she pay for all?

19. A man owes $85.08 for board, $12.50 for boots, and $58.70 for clothes: how much does he owe for all?

SUBTRACTION OF U. S. MONEY.

1. A man having $128.60, paid $47.735 for a cow: how much had he left?

ANALYSIS.—We write the less number under the greater, dollars under dollars, cents under cents, etc. Subtract in the usual way, and place the *decimal point* in the remainder under that in the subtrahend. (P. 112, Q. 5.)

Operation.
$128.60
47.735
————
Ans. $80.865

2. How subtract United States money?

Write the less number under the greater, dollars under dollars, cents under cents, etc., and subtract as in simple numbers, placing the decimal point in the remainder under that in the subtrahend.

NOTE.—If either of the given numbers has no *cents*, their place should be supplied by ciphers.

	(2.)	(3.)	(4.)	(5.)
From	$43.605	$107.38	$95.305	$100.00
Take	35.45	73.625	17.003	10.75

6. From $15.38 take $12.30.

7. From $110 take $68.40.

8. A man bought a horse for $250, and paid $118.50 down: how much does he still owe for it?

MULTIPLICATION OF U. S. MONEY.

1. What will 8 chairs cost, at $12.375 apiece?

ANALYSIS.—Since 1 chair costs $12.375, 8 chairs will cost 8 times $12.375. We multiply in the usual way, and from the right of the product point off *three* figures for cents and mills, because there are *three* places of cents and mills in the multiplicand.

Operation.
$12.375
8
———
Ans. $99.000

9. How multiply United States money?

Multiply as in simple numbers, and on the right of the product, point off as many figures for cents and mills as there are places of cents and mills in the multiplicand.

NOTE.—In *United States Money*, as in simple numbers, the *multiplier* must be considered an *abstract* number.

	(2.)	(3.)	(4.)	(5.)
Mult.	$25.40	$50.625	$95.45	$4350
By	8	9	15	24

	(6.)	(7.)	(8.)	(9.)
Mult.	$6207	$75.835	$463.05	$7584.25
By	45	52	62	75

10. What will 12 pounds of starch cost, at $0.125 a pound?

11. At $0.375 a pound, what will 25 pounds of butter come to?

12. What will 20 yards of cloth cost, at $6.75 a yard?

13. A miller sold 35 barrels of flour at $10.50 a barrel: what did it come to?

14. What cost 100 hats, at $5.50 apiece?

DIVISION OF U. S. MONEY.

1. A man bought 6 sleds for $21.42: what was that apiece?

ANALYSIS.—1 sled is 1 sixth of 6 sleds; therefore, 6)$21.42, 1 sled will cost 1 sixth of $21.42, which is $3.57.   *Ans.* $3.57

10. How divide U. S. money?

Divide as in simple numbers, and on the right of the quotient point off as many figures for cents and mills as there are places of cents and mills in the dividend.

NOTE.—For dividing money by money, or by a decimal divisor, the learner is referred to the Author's New Rudiments or New Practical Arithmetic.

2. Divide $25.65 by 5. 　　5. Divide $100.60 by 10.
3. Divide $36.48 by 8. 　　6. Divide $276 by 12.
4. Divide $63.72 by 9. 　　7. Divide $360 by 20.

8. A man sold 7 barrels of flour for $63.70: how much did he receive a barrel?

9. If you pay $25.74 for 6 yards of cloth, what will that be a yard?

10. If I pay $14.875 for 7 crates of peaches, how much will they cost me a crate?

11. A bookseller sold 5 slates for $0.685: what was that apiece?

12. If you divide $16.48 equally among 4 poor persons, how much will each receive?

13. If 8 yards of broadcloth cost $56.40, what will 1 yard cost?

14. If 7 pair of gloves cost $15.75, what is that a pair?

ENGLISH MONEY.

11. What are the denominations?

Pounds, shillings, pence, and farthings.

12. Recite the TABLE.

4 farthings (*qr.* or *far.*)	are	1 penny,	*d.*
12 pence	"	1 shilling,	*s.*
20 shillings	"	1 pound,	*£.*

NOTE.—The *pound sterling* is represented by a gold coin called a *sovereign,* and its value is $4.866½.

The value of an English *shilling* is 24¼ cents; and that of a *penny*, about 2 cents.

Sovereign.

TROY WEIGHT.

13. For what is Troy weight used?

For weighing *gold, silver*, and *jewels.*

14. What are the denominations?

Pounds, ounces, pennyweights, and grains.

15. Recite the TABLE.

24 grains (*gr.*)	are	1 pennyweight,	*pwt.*
20 pennyweights	"	1 ounce,	*oz.*
12 ounces	"	1 pound,	*lb.*

lb.   oz.   pwt.   gr.

1. How many ounces in 3 pounds? In 4 pounds?
2. In 40 pwt., how many ounces? In 42 pwt.?
3. In 24 ounces, how many pounds? In 39 ounces?

AVOIRDUPOIS WEIGHT.

16. For what is Avoirdupois Weight used?

For weighing all *coarse articles;* as, hay, groceries, etc., and all metals except *gold* and *silver*.

17. What are the denominations?

Tons, hundreds, pounds, and ounces.

18. Recite the TABLE.

16 ounces (*oz.*)	are	1 pound,	*lb.*
100 pounds	"	1 hundred weight,	*cwt.*
20 cwt., or 2000 lbs.,	"	1 ton,	*T.*

8 ounces are ½ pound.   4 ounces are ¼ pound.

Ton.   cwt.   lb.   oz.

NOTE.—The *ounce* is often divided into *halves, quarters,* etc., In business, the *dram,* the *quarter* of 25 lbs., and the *firkin* of 56 lbs., are not used as *units* of Avoirdupois Weight.

1. How many ounces in 2 pounds?   In 3 pounds?
2. How many ounces in 2 pounds and 5 ounces?
3. In 20 ounces, how many pounds?
4. In 32 ounces, how many pounds?   In 42 ounces?
5. In 40 hundred weight, how many tons?
6. How many hundreds in 3 tons?
7. How many hundreds in 5 tons?

APOTHECARIES' WEIGHT.

19. For what is Apothecaries' Weight used?

For *mixing medicines.*

20. What are the denominations?

Pounds, ounces, drams, scruples, and grains.

21. Recite the TABLE.

20 grains (*gr.*)	are 1 scruple, *sc.*, or ℈
3 scruples	" 1 dram, *dr.*, or ℨ.
8 drams	" 1 ounce, *oz.*, or ℥.
12 ounces	" 1 pound, ℔.

LINEAR MEASURE.

22. For what is Linear Measure used?

For measuring lines, distances, etc.

23. What are the denominations?

Leagues, miles, furlongs, rods, yards, feet, inches.

24. Recite the TABLE.

12 inches (*in.*)	are 1 foot,	*ft.*
3 feet	" 1 yard,	*yd.*
16½ feet, or 5½ yards	" 1 rod, perch, or pole, *r.* or *p.*
40 rods	' 1 furlong,	*fur.*
8 fur., or 320 rods	" 1 mile	*m.*
3 miles	" 1 league,	*l.*

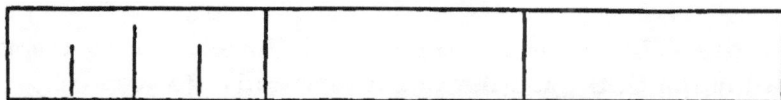

½ in.   1 in.   2 in.   3 in.

NOTE.—The inch is commonly divided into *halves, fourths, eighths,* or *tenths;* sometimes into *twelfths,* called *lines.*

1. George may draw a straight line 1 inch long upon the blackboard; the others on their slates.

2. Make one 2 in. long.   3 in. long.   6 in. long.

3. How long is your Arithmetic? How wide? How long is your slate?   How wide?

4. How long is this table?   How wide?

5. How many inches in 3 feet?   In 5 feet?

6. How many feet in 3 yards?   In 4 yards?

7. How many feet in 5 yards and 2 feet?

8. How many feet in 24 in.?   In 30 in.?   In 36 in.?

9. How many yards in 9 feet?   In 15 ft.?   In 24 ft.?

CLOTH MEASURE.

25. For what is Cloth Measure used?

For measuring cloths, laces, ribbons, etc.

26. What is the principal unit, and how divided?

The *linear yard*, which is divided into halves, fourths, eighths, etc.

27. Recite the TABLE.

3 ft., or 36 in.	are 1 yard,		yd.
18 in.	" 1 half yard,	$\frac{1}{2}$	yd.
9 in.	" 1 quarter yard,	$\frac{1}{4}$	yd.
4$\frac{1}{2}$ in.	" 1 eighth "	$\frac{1}{8}$	yd.
2$\frac{1}{4}$ in.	" 1 sixteenth "	$\frac{1}{16}$	yd.

NOTE.—*Ells* Flemish, English, and French, are no longer used in the U. S.; and the *nail* is practically obsolete.

1. How many yards in 8 half yards? In 17? In 24?

2. How many yards in 16 fourths?   In 28 fourths?

3. How many yards in 17 eighths?   In 34?

SQUARE MEASURE.

28. For what is Square Measure used?

For measuring *surfaces ;* as, land, flooring, etc.

29. What are the denominations?

Acres, square rods, square yards, square feet, and square inches.

30. Recite the TABLE.

144 square in. (*sq. in.*) are	1 square foot,	*sq. ft.*
9 square feet	" 1 square yard,	*sq. yd.*
30¼ square yards, or	" { 1 square rod,	*sq. r.*
272¼ square feet	{ perch, or pole,	
160 square rods	" 1 acre,	*A.*
640 acres	" 1 square mile,	*sq. m.*

1. A *square* is a rectilinear figure which has *four* equal sides, and *four* right angles.

2. A *square inch* is a square, each side of which is *one* inch in length.

3. A *square yard* is a square, each side of which is *one* yard in length.

NOTES.—1. The *corners* of any square figure, also of a book, a table, a room, etc., are *right angles.*

9 sq. ft. = 1 sq. yd.

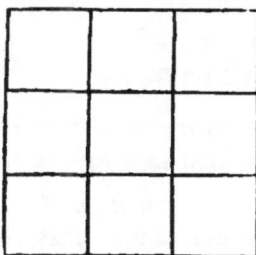

2. The acre was formerly divided into 4 roods : but, in practice, the rood is no longer used as a unit of measure.

1. Make a right angle upon the blackboard.

2. Make a square inch.

3. Make a square whose side is 3 inches. 6 inches. 8 inches.

4. Make a square foot. A square yard.

5. Divide a square yard into square feet.

6. Divide a square foot into square inches.

7. How many square feet in 5 square yards?

8. How many square yards in 36 square feet?

CUBIC MEASURE.

31. For what is Cubic Measure used?

For measuring *solids;* as, timber, boxes of goods, the capacity of rooms, ships, etc.

32. What are the denominations?

Cords, cubic yards, cubic feet, and cubic inches.

33. Recite the TABLE.

1728 cubic inches (*cu. in.*) are	1 cubic foot,	*cu. ft.*
27 cubic feet	" 1 cubic yard,	*cu. yd.*
128 cubic feet	" 1 cord,	*C.*

1. A *cube* is a regular solid bounded by *six equal squares,* called its faces.

12 × 12 × 12 = 1728.

2. A *cubic inch* is a cube, each side of which is a square inch.

3. A *cubic foot* is a cube, each side of which is a square foot.

4. A *cubic yard* is a cube, each side of which is a square yard.

3 ft. × 3 × 3 = 1 cu. yd.

5. A *Cord* of wood is a pile 8 ft. long. 4 ft. wide, and 4 ft. high; for, 8 × 4 × 4 = 128.

6. A *Cord* Foot = 16 cu. ft., and is 1 ft. long. 4 ft. wide, and 4 ft. high. 8 cord ft. make 1 cord.

NOTES.—1. Timber is now measured by cubic feet and inches.

2. The *old cubic ton* of 40 feet of round timber, or 50 feet of hewn timber, has fallen into disuse.

1. Draw a cubic inch.
2. Draw a cube whose sides are 2 inches square.
3. Draw a cubic foot.
4. How many inches long and wide must a block be to form a cubic foot?
5. How many feet high is a cubic yard?
6. How many cubic feet in 2 cubic yards?

DRY MEASURE.

34. For what is Dry Measure used?

For measuring *grain, fruit, salt,* etc.

35. What are the denominations?

Chaldrons, bushels, pecks, quarts, and pints.

36. Recite the TABLE.

2 pints (*pt.*)	are	1 quart,	*qt.*
8 quarts	"	1 peck,	*pk.*
4 pecks, or 32 qts.,	"	1 bushel,	*bu.*
36 bushels	"	1 chaldron,	*ch.*

bu.   ½ bu.   pk.   ½ pk.   qt.   pt.

NOTE.—The *dry* quart is equal to 1⅛ liquid quart nearly.

1. In 6 pints, how many quarts? In 12 pints?
2. In 12 quarts, how many pecks? In 16 quarts?

3. How many quarts in 3 pecks? In 7 pecks?
4. How many quarts in 5 pecks and 3 quarts?
5. In 5 pecks, how many bushels? In 8 pecks?
6. How many bushels in 12 pecks? In 15 pecks?
7. How many quarts in 2 bushels?
8. How many quarts in 1 bushel and 1 peck?
9. How many pints in 3 pecks?

LIQUID MEASURE.

37. For what is Liquid Measure used?

For measuring *milk, wine, molasses,* etc.

38. What are the denominations?

Hogsheads, barrels, gallons, quarts, pints, gills.

39. Recite the TABLE.

4 gills (*gi.*)	are	1 pint,	*pt.*
2 pints	"	1 quart,	*qt.*
4 quarts	"	1 gallon,	*gal.*
31½ gallons	"	1 barrel,	*bar.,* or *bbl.*
63 gallons, or 2 bbls.,	"	1 hogshead,	*hhd.*

hhd.   bar.   gal.   qt. pt. gi.

NOTES.—1. Liquid Measure is often called *Wine Measure.*
2. *Beer Measure* is practically obsolete in this country

1. How many gills in 2 pints? In 6 pints?
2. How many pints in 3 quarts? In 8 quarts?

3. How many quarts in 3 gallons? In 4 gallons?
4. How many quarts in 5 gallons and 2 quarts?
5. In 16 quarts, how many gallons?
6. In 8 pints, how many quarts? In 12 pints?
7. How many quarts in 13 pints? In 20 pints?
8. In 15 gills, how many pints? In 20 gills?
9. How many quarts in 6 gallons and 3 quarts?

CIRCULAR MEASURE.

40. For what is Circular Measure used?
For measuring *angles, latitude* and *longitude*, the *motion* of the heavenly bodies, etc.
41. What are the denominations?
Signs, degrees, minutes, and seconds.
42. Recite the TABLE.

60 seconds (″)	are	1 minute,	′
60 minutes	"	1 degree,	°
30 degrees	"	1 sign,	s
360 degrees	"	1 circumference, c or cir.	

NOTE.—Signs are used only in Astronomy.

A *Circle* is a plane figure bounded by a curve line, every part of which is *equally distant* from a point within called the *center*.

The adjoining figure is a circle. A D E B F is called the circumference; A B, the diameter; C A, C D, C E, etc., radii; and A D, D E, arcs.

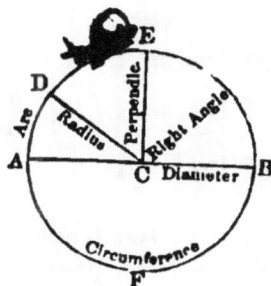

MISCELLANEOUS TABLES.

12 things are	1 dozen.		12 gross	"	1 great gro.
12 doz.	" 1 gross.		20 things	"	1 score.

24 sheets are	1 quire paper.		2 reams	are	1 bundle.
20 quires	" 1 ream.		5 bundles	"	1 bale.

2 leaves	are	1	folio.
4 leaves	"	1	quarto, or 4to.
8 leaves	"	1	octavo, or 8vo.
12 leaves	"	1	duodecimo, or 12 mo.
18 leaves	"	1	eighteen mo.
24 leaves	"	1	twenty-four mo.

NOTE.—The terms *folio*, *quarto*, *octavo*, etc., denote the number of leaves into which a sheet of paper is folded in making books.

4 in. =	1	hand, for measuring the height of horses.
9 in. =	1	span.
18 in. =	1	cubit.
6 ft. =	1	fathom, for measuring depths at sea.

32 lbs. =	1 bu. of oats.		58 lbs. =	1	bu. of corn.
48 lbs. =	1 " { buckwheat, or barley.		60 lbs. =	1	" { wheat, peas, or potatoes
56 lbs. =	1 " rye, or salt.		62 lbs. =	1	' beans.

100 pounds	=	1 quintal of dry fish.
196 pounds	=	1 barrel of flour.
200 pounds	=	1 barrel of fish, beef, or pork.
280 pounds	=	1 barrel of salt.

www.ingramcontent.com/pod-product-compliance
Lightning Source LLC
Chambersburg PA
CBHW021937190326
41519CB00009B/1052